HIGHER SCORES ON

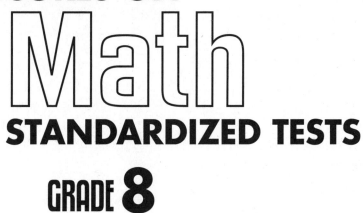

STANDARDIZED TESTS

GRADE 8

Contents

Standard	Descriptor	Pretest	Lessons	Practice Test A	Practice Test B
8.G.4	Understand that a two-dimensional figure is similar to another if the second can be obtained from the first by a sequence of rotations, reflections, translations, and dilations; given two similar two-dimensional figures, describe a sequence that exhibits the similarity between them.	38, 49	4, 6, 26, 39, 40, 46, 62, 76, 77, 83, 91, 105	6, 54	4, 29
8.G.5	Use informal arguments to establish facts about the angle sum and exterior angle of triangles, about the angles created when parallel lines are cut by a transversal, and the angle-angle criterion for similarity of triangles. *For example, arrange three copies of the same triangle so that the sum of the three angles appears to form a line, and give an argument in terms of transversals why this is so.*	39, 50	24, 32, 41, 45, 48, 59, 72, 95, 106, 110	1, 28	21, 46
	Understand and apply the Pythagorean Theorem.				
8.G.6	Explain a proof of the Pythagorean Theorem and its converse.				
8.G.7	Apply the Pythagorean Theorem to determine unknown side lengths in right triangles in real-world and mathematical problems in two and three dimensions.	40, 41	11, 19, 20, 29, 30, 42, 57, 58, 75, 84, 88, 89, 92	29, 32	7, 44
8.G.8	Apply the Pythagorean Theorem to find the distance between two points in a coordinate system.	42	14, 16, 17, 52, 54, 63, 82, 104, 109	42, 60	27, 36
8.G.9	Know the formulas for the volumes of cones, cylinders, and spheres and use them to solve real-world and mathematical problems.	43	3, 9, 12, 15, 21, 33, 68, 79, 85, 98, 107	19, 36	14, 22

Standard	Descriptor	Pretest	Lessons	Practice Test A	Practice Test B
	Statistics and Probability				
	Investigate patterns of association in bivariate data.				
8.SP.1	Construct and interpret scatter plots for bivariate measurement data to investigate patterns of association between two quantities. Describe patterns such as clustering, outliers, positive or negative association, linear association, and nonlinear association.	51, 55, 59	1, 5, 14, 20, 22, 24	27, 59	35, 42
8.SP.2	Know that straight lines are widely used to model relationships between two quantitative variables. For scatter plots that suggest a linear association, informally fit a straight line, and informally assess the model fit by judging the closeness of the data points to the line.	52, 56	2, 10, 19, 23	10, 45	12, 60
8.SP.3	Use the equation of a linear model to solve problems in the context of bivariate measurement data, interpreting the slope and intercept. *For example, in a linear model for a biology experiment, interpret a slope of 1.5 cm/hr as meaning that an additional hour of sunlight each day is associated with an additional 1.5 cm in mature plant height.*	53, 57	3, 6, 7, 11, 12, 13, 17, 21	5, 7	19, 49
8.SP.4	Understand that patterns of association can also be seen in bivariate categorical data by displaying frequencies and relative frequencies in a two-way table. Construct and interpret a two-way table summarizing data on two categorical variables collected from the same subjects. Use relative frequencies calculated for rows or columns to describe possible association between the two variables. *For example, collect data from students in your class on whether or not they have a curfew on school nights and whether or not they have assigned chores at home. Is there evidence that those who have a curfew also tend to have chores?*	54, 58, 60	4, 8, 9, 15, 16, 18	17	40, 52

Correlation Chart
Higher Scores on Math, Grade 8

Pretest

DIRECTIONS: Read each question and choose the best answer. Use the answer sheet provided at the end of the workbook to record your answers. If the correct answer is not available, mark the letter for "Not Here."

1. Look at the table below.

-3	$0.666...$	$\sqrt{5}$	9^3
a	b	c	d

Which one of the following is true for a, b, c, and d?

A

a	b	c	d
rational	irrational	irrational	rational

B

a	b	c	d
rational	rational	irrational	rational

C

a	b	c	d
irrational	irrational	irrational	rational

D

a	b	c	d
rational	irrational	rational	irrational

2. What is the best approximation for $\sqrt{5}$?

F between 2.3 and 2.4

G between 2.1 and 2.2

H between 1.5 and 1.6

J between 2.2 and 2.3

3. Which one of the following shows 25% as a fraction and as a decimal?

A $\frac{1}{2}$, 2.5

B $\frac{1}{25}$, 0.25

C $\frac{1}{4}$, 0.25

D $\frac{25}{100}$, 2.5

4. Which one of the following shows these numbers from least to greatest?

$$\sqrt{4}, -\sqrt{4}, 4^2, \text{ and } (\sqrt{4})^2$$

F $\sqrt{4}, -\sqrt{4}, 4^2, (\sqrt{4})^2$

G $-\sqrt{4}, 4^2, \sqrt{4}, (\sqrt{4})^2$

H $-\sqrt{4}, \sqrt{4}, (\sqrt{4})^2, 4^2$

J $(\sqrt{4})^2, -\sqrt{4}, 4^2, \sqrt{4}$

5. Simplify the expression.

$$(4^2)^6 \times \left(\frac{(6-2)^2}{4^5}\right) + (20-12)^3 \times 8^{10}$$

A $4^5 + 8^7$

B $4^9 + 8^{13}$

C $4^5 + 8^{30}$

D $4^5 + 8^{13}$

6. A square bathroom wall is covered in 81 square tiles. How many tiles are along each side of the wall?

F 9 tiles

G 18 tiles

H 36 tiles

J 72 tiles

7. What is 6,853,429,781 in scientific notation?

A $0.06853429781 \times 10^{11}$

B $6.853429781 \times 10^{9}$

C $6.853429781 \times 10^{-9}$

D 6853429781×10^{8}

8. The area of the United States is 9.629091×10^{6} km². The area of Texas is 6.95622×10^{5} km². How many km² larger is the United States than Texas?

F 2.672871×10^{-1} km²

G 2.672871×10^{1} km²

H 8.933469×10^{6} km²

J 8.933469×10^{11} km²

9. The graph shows Ann's running speed. What is her unit rate?

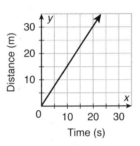

A 0 m/s

B 1.5 m/s

C 10 m/s

D 15 m/s

10. The growth of a plant is recorded in the table below.

Days	1	3	5	7	9
Height (in.)	4	10	16	22	a

What is the value of a? Does the table show linear growth?

F 28; yes

G 28; no

H 30; yes

J 30; no

11. Solve the equation.

$2^3x + 64 = (12 - 4)x - 5 + 8^3$

How many solutions does this equation have?

A   no solutions

C   more than one but less than 10 solutions

B   one solution

D   an infinite number of solutions

---

12. Solve the equation.

$3.2(x + 4) + 2^2 + 1.7(x - 3) = 27$

F $x = 3.12$ **G** $x = 4.48$ **H** $x = 4.89$ **J** $x = 2.2$

---

13. Which of the graphs below is a solution to this system of equations?

$2x - 3y = -2$
$4x + y = 6$

A
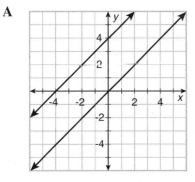

C
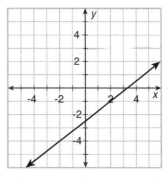

B
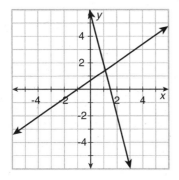

D
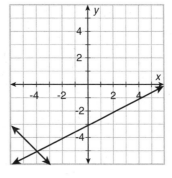

14. Which ordered pair is a solution to this system of equations?

$2x + y = 9$
$3x - y = 16$

F $(-5, -1)$ **G** $(-5, 1)$ **H** $(5, -1)$ **J** $(1, -5)$

---

15. The local middle school had a concert to raise money. Tickets were $6.00 for adults and $4.00 for children. The committee figured there were 480 in attendance and that $2,340.00 was raised. How many adults and how many children attended?

A 45 adults and 235 children

B 255 adults and 225 children

C 210 adults and 270 children

D 270 adults and 210 children

---

16. Look at the graph and the table below.

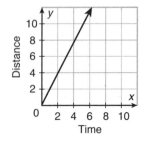

Time	1.5	2.5	3.5	4.5
Distance	4.5	7.5	10.5	13.5

Figure A   **Figure B**

Which table correctly compares the rates of speed shown in Figures A and B above?

F

Figure A	Figure B	Rate of Figures A and B
$\frac{1}{2}$	$\frac{1}{2}$	A and B are the same.

H

Figure A	Figure B	Rate of Figures A and B
2	3	B is faster.

G

Figure A	Figure B	Rate of Figures A and B
2	3	A is faster.

J

Figure A	Figure B	Rate of Figures A and B
3	2	A is faster.

17. If the slope of a line is $\frac{3}{4}$, which two points are on the line?

   A $(1, 3)$ and $(-1, 2)$ **B** $(-1, -1)$ and $(3, 3)$ **C** $(5, 2)$ and $(3, 2)$ **D** $(-1, -1)$ and $(3, 2)$

---

18. Solve for x.

$$8x + 9 - 2x = 6 + 6x + 3$$

Which one of the following best describes the solution for the equation?

   F no solution   **G** all real numbers   **H** $x = 6$ **J** $x = 5$ and $x = 15$

---

19. Solve the equation.

$$13 - (2x + 2) = 2(x + 2) + 3x$$

   A $x = 1$ **B** $x = 2$ **C** $x = 3$ **D** $x = 7$

---

20. Which graph shows a system of equations with no solution?

F
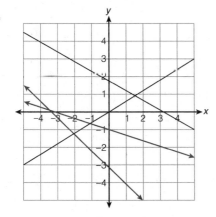

H
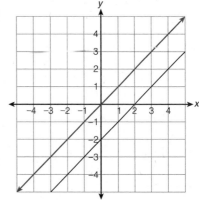

G
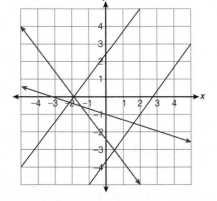

J
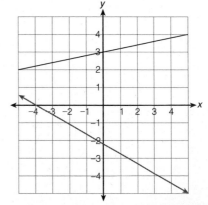

21. Solve for x and y in this system of equations.

$2y = 3x - 2$
$y = 2x + 1$

A $x = -7, y = -4$

B $x = -4, y = -7$

C $x = 4, y = 7$

D $x = 7, y = 4$

22. The seventh grade basketball team scores 80 points when they play against the eighth grade team. They make a total of 37 baskets, some worth 2 points and some worth 3 points. How many of each do they make?

F

2 points	3 points
6	31

G

2 points	3 points
5	32

H

2 points	3 points
32	5

J

2 points	3 points
31	6

23. Which group of ordered pairs does NOT define a function?

A $(1, 5), (2, 6), (2, 7), (3, -8)$

B $(1, 4), (2, 8), (3, 12), (4, 16)$

C $(0, 0), (2, 4), (3, 6), (4, 8)$

D $(-1, 3), (2, 6), (3, 9), (4, 8)$

24. Look at the two tables below.

Time	1	2	3	4	5
Distance	4	8	12	16	20

Figure A

Time	1	2	3	4	5
Distance	3	6	9	12	15

Figure B

What is the rate of change in each table, and which is faster?

F Figure A: 4, Figure B: 3, A is faster

G Figure A: 4, Figure B: 3, B is faster

H Figure A: 3, Figure B: 4, B is faster

J Figure A: 3, Figure B: 4, A is faster

25. Look at the graph below.

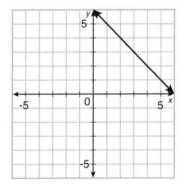

Which one of the following does NOT describe the function?

A $y = -x + 6$

B $4y = \frac{1}{4}x + 18$

C $y = 6 - x$

D $3y = -3x + 18$

26. David gets paid $150.00 per week, plus $5.00 for every window he washes. Which equation shows the relationship between the number of windows he washes and his total weekly income?

   F $y = 150x + 5$

   G $2y = 300x + 10$

   H $y = 5x + 150$

   J $\frac{1}{3}y = 50x + 1.6$

27. Helena is making a graph of her progress with math homework. When she starts, she is able to work just a few problems each day. However, as time goes on, the subject becomes harder, and she has to work more problems each day to learn the material. Which graph represents her math study?

   A
   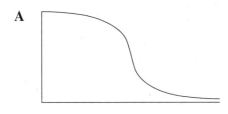

   B
   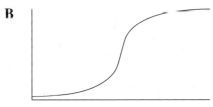

   C
   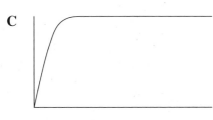

   D
   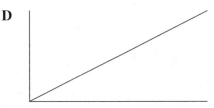

28. Look at the table below.

−4	−2	1	2	3
2	3	a	5	5.5

   What is the value of a so that the table defines a linear function?

   F 3.25   **H** 4.5

   G 3.5   **J** 4.75

29. Look at the table and the graph below.

x	−2	−4	0	5
y	−5	−5	−5	−5

   Figure A

   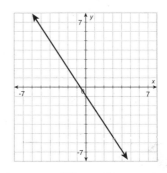

   Figure B

   What are the slopes in Figures A and B, and which one has the greater rate of change?

   A Slope in Figure A is 0. Slope in Figure B is $-\frac{3}{2}$. Figure B has a greater rate of change.

   B Slope in Figure A is 0. Slope in Figure B is $-\frac{3}{2}$. Figure A has a greater rate of change.

   C Slope in Figure A is 0. Slope in Figure B is $-\frac{2}{3}$. Figure B has a greater rate of change.

   D Slope in Figure A is 0. Slope in Figure B is $-\frac{2}{3}$. Figure A has a greater rate of change.

30. Jacob is driving to San Antonio, Texas. He records his progress in the table below.

Time (hr)	3	5	7	9
Distance (mi)	210	350	490	a

How many miles will he have driven after 9 hours of driving?

F 350   **H** 660

G 630   **J** 882

31. This equation shows how the total distance May has ridden her bicycle depends on the number of trips she makes to work. The total distance she has ridden in miles is represented by d. The number of trips she has made is represented by t.

$$d = 2t + 50$$

How many trips will she have to make to ride a total of 100 miles?

A 25   **C** 75

B 50   **D** 250

32. The graph below shows the yearly sales for the L&T Corporation.

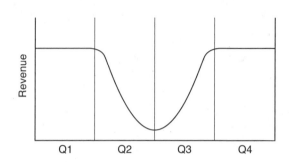

Which one of the following describes the sales?

F Sales were flat in Q1, dropped in Q2 and Q3, and increased in Q4.

G Sales were about the same in Q1 and Q4. In Q2 and Q3, sales increased.

H Sales were steady and high in Q1 but dropped in Q2. In Q3, sales increased and held steady in Q4.

J Sales in Q2 and Q3 were higher than in Q1 and Q4.

33. The vertices of triangle ABC on a coordinate plane are $C = (-4, 2)$, $B = (-8, 2)$, and $A = (-8, 7)$. If the triangle is translated so its vertices are $C' = (-2, 4)$, $B' = (-6, 4)$, and $A' = (-6, 9)$, which statement is true about the sides of the triangle?

A $CB > B'C'$

B $AC < A'C'$

C $AB + BC \neq A'B' + B'C'$

D $AB + BC + CA = A'B' + B'C' + C'A'$

34. The vertices of trapezoid *ABCD* are at points
A = (−8, −3), *B* = (−9, −6), *C* = (−2, −6),
and *D* = (−3, −3). The trapezoid is translated
up 10. Which one of the following is NOT true
about trapezoids *ABCD* and *A′B′C′D′*?

F $\angle ABC$ and $\angle A'D'C'$ are supplementary
angles.

G $\angle ABC = \angle B'C'D'$

H $\angle BAD = \angle C'D'A'$

J $\angle BAD$ and $\angle C'D'A'$ are complementary.

35. Two parallel lines drawn on a coordinate
plane each have a slope of 3. Which one of the
following is true about the reflection over the
x-axis of the lines?

A   The reflected lines are closer to each other
than the original lines.

B   The slope of the reflected lines is −3.

C   The slope of the reflected lines is $\frac{1}{3}$.

D   The slope of the reflected lines is 3.

36. Look at the graph below.

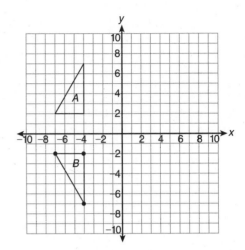

Which one of the following transformations was
done to figure A to transform it into figure B?

F   Figure A was translated to figure B.

G   Figure A was reflected over the *x*-axis to
form figure B.

H   Figure A was rotated around the origin to
form figure B.

J   Figure B is a dilation of figure A.

37. Look at the graph below.

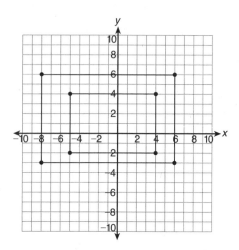

Which one of the following is the scale factor that transforms the smaller rectangle into the larger one?

A   1   **C**   2

B   1.5   **D**   2.5

38. Look at the graph below.

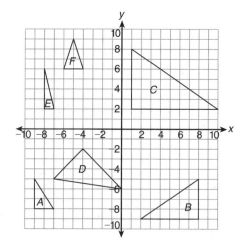

Which figures are similar to one another?

F   figure A, figure B, and figure C

G   figure B, figure C, and figure D

H   figure A, figure F, and figure C

J   figure D, figure E, and figure A

39. Look at the figure below.

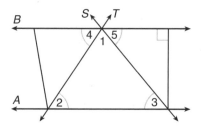

Which one of the following is NOT true if $A \parallel B$?

A $\angle 2 = \angle 4$

B $\angle 2 + \angle 5 + \angle 1 = 180°$

C $\angle 4 + \angle 5 = 90°$

D $\angle 3 = \angle 5$

40. What is the length of the side of a cone if the diameter of the base is 10 and the height is 11?

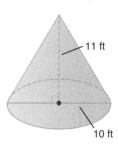

11 ft

10 ft

F   4   **H** $\sqrt{96}$

G $\sqrt{146}$ **J**   12

41. The table below gives the lengths of the three sides of four triangles.

Triangle	Lengths
A	9, 12, 14
B	10, 24, 25
C	7, 8, 9
D	5, 12, 13

Which triangle is a right triangle?

A △A

C △C

B △B

D △D

42. Look at the graph below.

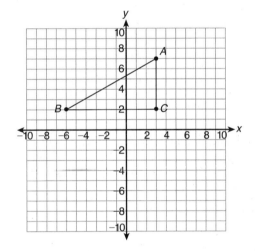

Which one of the following is the best estimate of the length of side AB of the right triangle shown?

F between 10 and 11

G between 10 and 10.5

H between 10 and 10.3

J between 10 and 10.1

43. What is the volume of a cone if the radius of its base is 7 inches and its height is 10 inches? Use $\pi = 3.14$.

A $V = 146$ in.^{3}

B $V = 512.9$ in.^{3}

C $V = 769$ in.^{3}

D Not Here

44. If rectangle ABCD has its vertices on points $A = (1, -3)$, $B = (1, -6)$, $C = (6, -6)$, and $D = (6, -3)$ and is translated 3 to the right and 3 up, which one of the following is true about rectangle $A'B'C'D'$?

F The perimeter of $A'B'C'D'$ is 3 times larger than that of ABCD.

G The perimeter of ABCD equals the perimeter of $A'B'C'D'$.

H The length of AB is greater than the length of BC.

J ABCD is not congruent to $A'B'C'D'$.

45. If triangle ABC is drawn on a coordinate plane and is translated 3 to the left and 2 down, which one of the following is a true statement about triangle ABC and triangle $A'B'C'$?

A The point (0, 0) will fall inside the second triangle.

B $\angle A$ and $\angle A'$, $\angle B$ and $\angle B'$, and $\angle C$ and $\angle C'$ are supplementary.

C $\angle A = \angle A'$, $\angle B = \angle B'$, and $\angle C = \angle C'$.

D The triangles are similar but are not congruent.

46. Rectangle *ABCD* is translated to *A′B′C′D′*. Which one of the following is NOT true about the sides of the two rectangles?

F *AB* ∥ *BC* and *A′B′* ∦ *C′D′*

G *AD* ∥ *BC* and *A′D′* ∥ *B′C′*

H *AD* ∦ *CD* and *A′B′* ∦ *C′D′*

J Not Here

47. Look at the graph below.

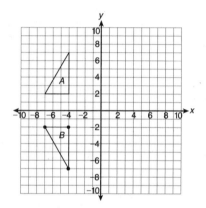

Which one of the following is true of figure A and figure B?

A Figure A is congruent to figure B.

B Figure A and figure B are not congruent, but they are similar.

C The area of figure B is larger than the area of figure A.

D The perimeter of figure B is not equal to the perimeter of figure A.

48. Look at the graphs below. The figure in Graph A was dilated to form the figure in Graph B.

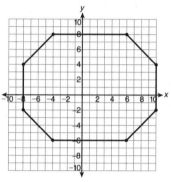

Graph A

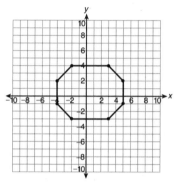

Graph B

What is the scale factor from Graph A to Graph B?

F 0.5   **H** 2

G 1   **J** 2.5

49. Look at the figure below.

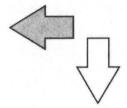

Which transformation will move the shaded figure onto the unshaded figure?

A   translation followed by rotation   **C**   translation followed by reflection

B   rotation followed by dilation   **D**   reflection followed by dilation

---

50. Look at the figure below.

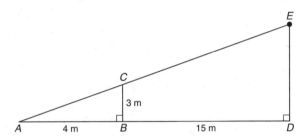

Which proportion can be used to find the length of *DE*?

F $\dfrac{CB}{AB} = \dfrac{DE}{BD}$ **G** $\dfrac{CB}{AB} = \dfrac{ED}{AD}$ **H** $\dfrac{AC}{AB} = \dfrac{AE}{AD}$ **J** $\dfrac{CE}{BD} = \dfrac{CB}{ED}$

---

51. Look at the graph below.

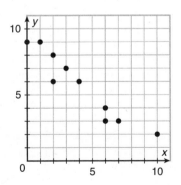

Which one of the following describes the graph?

A   cluster   **B**   no association   **C**   negative association   **D**   outlier

52. In which one of the following is a trend line most helpful?

   F   linear association   **G**   cluster   **H**   no association   **J**   nonlinear association

---

53. Look at the scatter plot below.

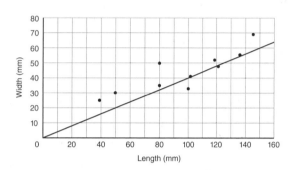

Which one of the following gives the slope and equation of the trend line?

A   slope: $\frac{5}{2}$; equation: $y = \frac{5}{2}x$ **C**   slope: $\frac{2}{5}$; equation: $y = \frac{5}{2}x$

B   slope: $\frac{2}{5}$; equation: $y = \frac{2}{5}x$ **D**   slope: $\frac{5}{2}$; equation: $y = \frac{2}{5}x$

---

54. George is collecting information about pets owned by his classmates. He wants to find out how many of the 30 students in the class have a dog and how many have a cat. He records the information in the table below.

	Dog	**No Dog**
Cat	cat and dog	a
No Cat	b	c

Which one of the following gives the correct replacements for a, b, and c in the table?

F

a	b	c
dog, no cat	cat, no dog	no dog and no cat

H

a	b	c
cat, no dog	dog, no cat	no dog and no cat

G

a	b	c
no dog and no cat	cat, no dog	dog, no cat

J

a	b	c
cat, no dog	no dog and no cat	dog, no cat

55. Look at the table below.

Price ($)	3	6	9	12	18	21
Buyers	30	22	15	2	1	1

Which scatter plot is the best representation of the data?

A
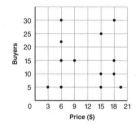

B
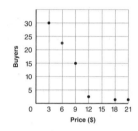

C
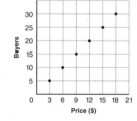

D

56. The scatter plot below compares the length and width of people's hands.

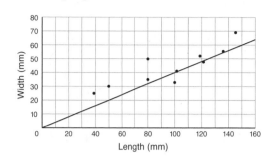

Which one of the following is the best guess for the width when the length is 65 mm?

F   22   H   29

G   25   J   30

57. Look at the scatter plot below.

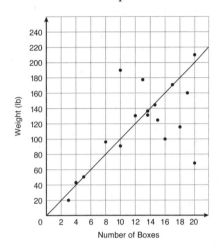

Number of Boxes

Which one of the following is the best description of the scatter plot?

A positive linear association; slope 10; equation $y = 10x$

B positive linear association; slope $\frac{1}{10}$; equation $y = \frac{1}{10}x$

C negative linear association; slope $\frac{1}{10}$; equation $y = \frac{1}{10}x$

D negative linear association; slope 10; equation $y = 10x$

58. Juanita asks 150 people if they prefer chocolate or strawberry ice cream. She records the results in the table below.

	Chocolate	Strawberry	Total
Men	50	20	70
Women	35	45	80
Total	85	65	150

What is the relative frequency of men who like chocolate?

F 13%   **H** 46%

G 33%   **J** 56%

59. What are data that involve two variables called?

A positive association data

B negative association data

C scatter plot data

D bivariate data

60. Look at the table below.

	Chocolate	Strawberry	Total
Men	50	20	70
Women	35	45	80
Total	85	65	150

Comparing relative frequency, which one of the following is the most likely association?

F men who like chocolate and women who like strawberry

G men who like chocolate and women who like chocolate

H men who like chocolate and men who like strawberry

J women who like chocolate and men who like strawberry

Name _____   Date _____

The Number System

Modeled Instruction

DIRECTIONS: Read each question and choose the best answer. Use the answer sheet provided at the end of the workbook to record your answers. If the correct answer is not available, mark the letter for "Not Here."

1. Look at the table below containing some rational and irrational numbers.

-25	0.875	π	$\sqrt{2}$
a	b	c	d

Which one of the following shows the correct classification of a, b, c, and d?

A

a	b	c	d
rational	irrational	irrational	irrational

B

a	b	c	d
rational	rational	irrational	irrational

C

a	b	c	d
rational	irrational	irrational	rational

D

a	b	c	d
irrational	rational	irrational	irrational

Hint

A rational number can be written as the ratio of two integers.

2. Which is the correct number line to show the numbers $\sqrt{10}$, $-\sqrt{10}$, $(\sqrt{10})^2$?

F

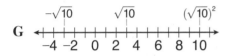

G

H

J

Hint

Find the approximate decimal value of each number.

3. Which one of the following is NOT a characteristic of the irrational number π?

A π is the ratio of a circle's circumference to its diameter.

B π cannot be expressed as the ratio of two integers.

C π can accurately be written as the fraction $\frac{22}{7}$.

D   The decimal representation of π contains an infinite number of non-repeating digits.

Hint

We commonly use an approximation of π.

4. Which one of the following is the closest approximation for $\sqrt{2}$?

F   between 1.39 and 1.5

G   between 1 and 4

H   between 1 and 2

J   between 1.39 and 1.42

Hint

$\sqrt{2}$ is approximately 1.4142…

5. Which one of the following shows the fraction and decimal equivalents of $33\frac{1}{3}\%$?

A $\frac{1}{3}$, 0.333…   C $\frac{3}{100}$, 3.3

B $\frac{1}{3}$, 0.33   D $\frac{1}{3}$, $33\frac{1}{3}$

Hint

Percent means out of 100.

6. Which one of the following shows these numbers in order from least to greatest: $\sqrt{25}$, π, $\sqrt{4}$, $\sqrt{2}$, $\sqrt{7}$?

F $\sqrt{2}$, $\sqrt{4}$, $\sqrt{7}$, π, $\sqrt{25}$

G $\sqrt{2}$, $\sqrt{4}$, π, $\sqrt{7}$, $\sqrt{25}$

H $\sqrt{2}$, π, $\sqrt{4}$, $\sqrt{7}$, $\sqrt{25}$

J $\sqrt{25}$, π, $\sqrt{7}$, $\sqrt{4}$, $\sqrt{2}$

Hint

Find the approximate decimal equivalents.

The Number System

Independent Practice

DIRECTIONS: Read each question and choose the best answer. Use the answer sheet provided at the end of the workbook to record your answers. If the correct answer is not available, mark the letter for "Not Here."

7. Which one of the following shows these numbers in order from least to greatest: $\sqrt{7}, \pi, -\frac{1}{3}, \sqrt{11}$?

   A $-\frac{1}{3}, \pi, \sqrt{7}, \sqrt{11}$ **B** $-\frac{1}{3}, \sqrt{11}, \pi, \sqrt{7}$ **C** $-\frac{1}{3}, \sqrt{11}, \sqrt{7}, \pi$ **D** $-\frac{1}{3}, \sqrt{7}, \pi, \sqrt{11}$

8. Which one of the following is the closest approximation for $\sqrt{13}$?

   F between 3 and 4   **G** between 3.59 and 3.65   **H** between 3.5 and 4   **J** between 3 and 3.9

9. Which one of the following is true for the expressions $\sqrt{14} + 5$ and $\sqrt{16} + 9$?

   A $\sqrt{14} + 5 > \sqrt{16} + 9$ **C** $\sqrt{14} + 5 < \sqrt{16} + 9$

   B $\sqrt{14} + 5 \geq \sqrt{16} + 9$ **D** $\sqrt{14} + 5 = \sqrt{16} + 9$

10. Look at the table below containing some rational and irrational numbers.

3.333...	$\sqrt{9}$	0.25	$\sqrt{7}$	11
a	b	c	d	e

Which one of the following shows the correct classifications for a, b, c, d, and e?

F

a	b	c	d	e
rational	rational	rational	irrational	rational

G

a	b	c	d	e
irrational	irrational	rational	rational	irrational

H

a	b	c	d	e
irrational	rational	irrational	irrational	rational

J

a	b	c	d	e
irrational	rational	rational	irrational	irrational

11. Which one of the following shows these numbers ordered from least to greatest?

$\pi, \sqrt{16}, \sqrt{7}, \sqrt{9}, 5$

A $\sqrt{7}, \sqrt{9}, \pi, \sqrt{16}, 5$

B $\sqrt{7}, \pi, \sqrt{9}, \sqrt{16}, 5$

C $\sqrt{9}, \sqrt{7}, \pi, \sqrt{16}, 5$

D $5, \sqrt{16}, \pi, \sqrt{9}, \sqrt{7}$

12. Which one of the following is a characteristic of an irrational number?

F It has a repeating decimal.

G It is the square root of a perfect square.

H It cannot be expressed as a fraction.

J It is a terminating decimal.

13. Which one of the following is an example of an irrational number?

A $\sqrt{13}$ C $\sqrt{25}$

B $\sqrt{16}$ D $\sqrt{100}$

14. Which one of the following is the closest approximation for $\sqrt{113}$?

F between 10.5 and 10.7

G between 10 and 11

H between 10 and 10.7

J between 10.5 and 10.8

15. Which one of the following is the correct comparison of the numbers: $2\sqrt{13} + 4\sqrt{17}$ and $\pi - 2\sqrt{2} + 5\sqrt{3}$?

A $2\sqrt{13} + 4\sqrt{17} \geq \pi - 2\sqrt{2} + 5\sqrt{3}$

B $2\sqrt{13} + 4\sqrt{17} = \pi - 2\sqrt{2} + 5\sqrt{3}$

C $2\sqrt{13} + 4\sqrt{17} > \pi - 2\sqrt{2} + 5\sqrt{3}$

D $2\sqrt{13} + 4\sqrt{17} < \pi - 2\sqrt{2} + 5\sqrt{3}$

16. Which one of the following shows these numbers ordered from least to greatest?

$3.5, \pi, \sqrt{7}, 8, \sqrt[3]{8}$

F $\sqrt{7}, \sqrt[3]{8}, \pi, 3.5, 8$

G $\sqrt[3]{8}, \sqrt{7}, \pi, 3.5, 8$

H $\sqrt[3]{8}, \sqrt{7}, 3.5, \pi, 8$

J $\sqrt{7}, \sqrt[3]{8}, 3.5, \pi, 8$

17. Which one of the following numbers appears to be irrational?

A 5.25×10^{-8}

B $2.718281828459045235360287413 52...$

C $4.666666666666666666666666666 666...$

D $\sqrt{1,000,000}$

18. An average ostrich egg is 10.159 centimeters by 17.78 centimeters. What are these measurements as fractions in simplest form?

F $\dfrac{1,000}{10,159}$ by $\dfrac{100}{889}$ cm

G $\dfrac{10,159}{1,000}$ by $\dfrac{889}{100}$ cm

H $10\dfrac{159}{1,000}$ by $17\dfrac{39}{50}$ cm

J $10\dfrac{1,000}{159}$ by $17\dfrac{50}{39}$ cm

19. Which one of the following is true for the numbers 2.5 and $2\frac{4}{5}$?

A $2.5 > 2\frac{4}{5}$

C $2.5 < 2\frac{4}{5}$

B $2.5 = 2\frac{4}{5}$

D $2.5 \geq 2\frac{4}{5}$

20. Which one of the following is NOT a characteristic of a rational number?

F It can be expressed as a fraction.

G It cannot be expressed as a fraction.

H It can be a repeating decimal.

J It can be a terminating decimal.

Expressions and Equations

Modeled Instruction

DIRECTIONS: Read each question and choose the best answer. Use the answer sheet provided at the end of the workbook to record your answers. If the correct answer is not available, mark the letter for "Not Here."

1. The graph shows Tony's speed.

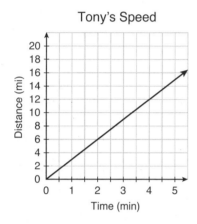

Tony's Speed

What is his rate of change?

A 0.75

C 3

B 1

D 6

Hint

Look at the value of y when $x = 1$.

2. Evaluate the expression for the given value of the variables.

$c^3 + d^3$ when $c = 2$ and $d = 5$

F 21

H 133

G 29

J 343

Hint

Substitute the numbers for the variables.

3. A cube has a volume of 63 ft³. What is the approximate length of each side in inches?

A 0.333… in.

C 4 ft

B 0.444… in.

D 48 in.

Hint

The volume of a cube is s^3.

4. What is the slope of the equation $7x + 4y = 24$?

F $-\dfrac{7}{4}$

H 4

G $-\dfrac{4}{7}$

J 6

Hint

The slope-intercept form of an equation is $y = mx + b$.

Name _____ Date _____

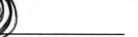

5. Look at the graph below.

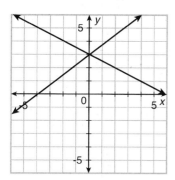

Which ordered pair represents the solution to the system of equations represented by the lines on the graph?

A $(2, 2)$ B $(-4, 0)$ C $(0, 3)$ D $(3, 0)$

Hint

The point of intersection of the two lines is the solution to the equations.

6. Jamie bought some bats and baseballs. She bought a total of 20 items and spent $110.00. The bats (x) cost $5.00 each, and the baseballs (y) cost $7.50 each. How many of each did she buy?

F   17 bats and 3 balls

G   16 balls and 4 bats

H   16 bats and 4 balls

J   3 bats and 17 balls

Hint

Write two equations, one representing the number of items and one representing the cost.

7. Two times a number is equal to six less than 3 times the number. What is the number?

A -6 C $\dfrac{3}{2}$

B $\dfrac{2}{3}$ D 6

Hint

Let x equal the number. Write an equation and solve for x.

8. When an equation is solved, the result is $5 = 15$. How many solutions are there for the equation?

F 0   H 2

G 1   J infinite number

Hint

What is the consequence of $5 \neq 15$?

9. The school band gave a concert. Adult tickets were $5.00, and student tickets were $3.00. There were 100 people in the audience, and the band made $500.00 in ticket sales. Which system of equations can be used to find out how many of each kind of tickets were sold?

A $8(a + s) = 500$ and $a + s = 100$

B $a + s = 100$ and $5a + 3s = 500$

C $a + s = 500$ and $8(a + s) = 100$

D $a + s = 500$ and $5a + 3s = 100$

Hint

Write an equation for the number of people and an equation for the amount raised.

Expressions and Equations
Higher Scores on Math, Grade 8

10. The mass of the sun is approximately 2.00×10^{30} kilograms. The mass of the moon is approximately 7.36×10^{22} kilograms. The mass of the sun is approximately how many times the mass of the moon? The answer should be in scientific notation.

F   27,000,000   H 3.69×10^{-8}

G 2.7×10^7 J   0.0000000369

Hint

Set up a ratio of the sun's mass to the moon's mass.

11. If the slope of a line is $\frac{5}{6}$, which two points are on the line?

A   (6, 12) and (5, 10)

B   (12, 6) and (10, 5)

C   (6, 5) and (12, 10)

D   (10, 12) and (5, 6)

Hint

Use the equation $y = mx$.

12. 2.6645×10^5 is the scientific notation of what number?

F   0.26645   H   266,450

G   26,645   J   2,664,500

Hint

Move the decimal point to the right the number of places specified by the exponent.

13. Which graph shows a system of equations with an infinite number of solutions?

A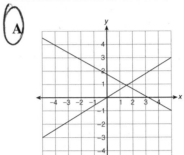

B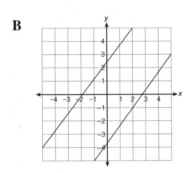

C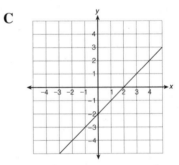

D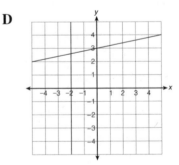

Hint

The solution to the equations is where the two lines intersect.

14. Look at the data table below.

x	−2	−1	0	1	2
y	5	3	1	−1	−3

Which one of the following graphs represents the data in the table?

F

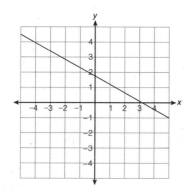

H

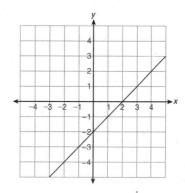

G

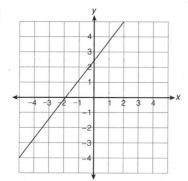

J

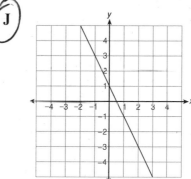

Hint

Look at each graph to find the one in which the line goes through the points in the table.

---

15. Which one of the following is the slope of the line passing through points $(−2, 3)$ and $(4, 6)$?

A $−2$ **B** $−\dfrac{1}{2}$ **C** $\dfrac{1}{2}$ **D** 2

Hint

Slope equals $\dfrac{y_2 − y_1}{x_2 − x_1}$.

16. Look at this system of equations.

$$y = 6x + 5$$
$$3y = 18x + 15$$

Which one of the following will help solve the system of equations?

F $3y = 18\left(\dfrac{y+5}{6}\right) + 15$

G $3(6x + 5) = 18x + 15$

H $3y = 18(6x + 5) + 15$

J $18x + \dfrac{15}{3} = 6x + 5$

Hint

Substitute the value of y from the first equation for y in the second equation.

17. Solve the equation.

$$2x + 5 - 3x = 19x + 2(25 - 10x)$$

How many solutions are there for this equation?

A 0   **C** 2

B 1   **D** infinite number

Hint

Solve for x using the properties of equations.

18. Solve the equation.
$$5(2 - x) - 3(4 - 2x) = 20$$

F $x = -22$

G There is no solution.

H There are an infinite number of solutions.

J $x = 22$

Hint

Solve for x using the properties of equations.

19. What value of x makes the equation true?

$$3(5 - x) - 2(5 + x) = 3(x + 1)$$

A $\dfrac{1}{4}$ **C** 4

B 2   **D** $\dfrac{4}{3}$

Hint

Multiply to clear the parentheses and then solve.

20. Solve the equation.

$$11 + (3x - 7) = (6x + 5) - 3x$$

How many solutions are there for this equation?

F   no solutions

G   one solution

H   more than one solution, but less than 10 solutions

J   an infinite number of solutions

Hint

Solve for x by combining like terms.

21. Which one of the following is the scientific notation form of 135,677?

A 1.35×10^{-5} **C** 13.5×10^5

B 1.35×10^5 **D** 0.135×10^{-5}

Hint

The number must be between 1 and 10 times a power of 10.

22. Which of the data tables below gives solutions to the equation $y = x + (3x + 5)$?

F

x	1	2	3	4
y	9	13	17	21

G

x	3	4	5	6
y	17	21	25	28

H

x	2	3	4	5
y	13	18	21	25

J

x	7	8	9	10
y	33	37	43	47

Hint

Substitute the values for x and solve for y.

23. Look at the graph and table below.

Figure A

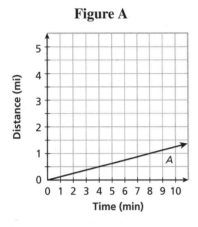

Figure B

Time (min)	0.15	1	1.5	2	2.5
Distance (mi)	0.1	0.2	0.3	0.4	0.5

What are the rates of change shown in the graph and table, and which shows the faster speed?

A

Figure A	Figure B	Speed
$\frac{1}{5}$	$\frac{1}{8}$	B is faster than A.

B

Figure A	Figure B	Speed
$\frac{1}{8}$	$\frac{1}{5}$	A is faster than B.

C

Figure A	Figure B	Speed
$\frac{1}{5}$	$\frac{1}{8}$	A is faster than B.

D

Figure A	Figure B	Speed
$\frac{1}{8}$	$\frac{1}{5}$	B is faster than A.

Hint

Determine the rate of change (equal to the slope) either by reading the graph or dividing numbers in the table.

24. Solve the equation.

$$4x + 3y - 5 = 28 + (4x + 3y)$$

How many solutions are there for this equation?

F There are an infinite number of solutions.

G There is one solution.

H There are no solutions.

J There are exactly two solutions.

Hint

Combine like terms and solve for x.

25. How many boxes 3 cm × 3 cm × 3 cm will fit into a carton that is 9 cm on each side?

A 3   **C** 72

B 27   **D** 702

Hint

Find the volume of the boxes and the volume of the carton. $V = s^3$.

26. Evaluate the expression.

$$(3^3)^3 + ((7 + 5)^2 \times 4^{-2}) + (15 - 103) + 25$$

F 19,759   **H** 19,568

G 19,629   **J** 19,479

Hint

Powers of exponents are multiplied. Negative exponents mean $\frac{1}{n}$. Order of operations must be used.

Name _____   **Date** _____

27. What is $(2.0 \times 10^{-1})^3$ in scientific notation?

A 2.0×10^{-3} C 8.0×10^{-3}

B $2.0 \times \dfrac{1}{10^3}$ D $8.0 \times \dfrac{1}{10^3}$

Hint

Use the rule for raising to a power. Scientific notation means the number is between 1 and 10 times a power of ten.

28. The Pacific Ocean covers about 1.66241×10^8 square kilometers. The Baltic Sea covers about 4.144×10^5 square kilometers. The Pacific Ocean is approximately how many times larger than the Baltic Sea?

F 0.40057×10^3 H 4.0×10^2

G 2.49278×10^{-3} J 2.4927278×10^3

Hint

Set up a ratio to compare the two bodies of water.

29. A rocket travels at the speed of light, 3×10^8 meters per second. How many seconds will it take to reach the moon when it is 3.844×10^8 kilometers from Earth?

A 0.78043

B 7.8043×10^{-1}

C $1.281333... \times 10^0$

D 12.8133×10^1

Hint

Divide the distance by the speed and change to scientific notation.

30. Look at the graph below that shows the amount of flour needed to bake a specific number of cupcakes.

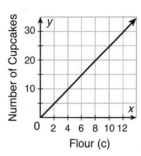

What is the slope of the line in the graph?

F $\dfrac{2}{5}$ H $\dfrac{5}{2}$

G $\dfrac{14}{35}$ J Not Here

Hint

Slope equals $\dfrac{y_2 - y_1}{x_2 - x_1}$.

31. What is 0.00088 written in scientific notation?

A 8.8×10^{-4} C 88×10^{-5}

B 8.8×10^4 D 88×10^5

Hint

The number must be between 1 and 10 times a power of ten.

32. Evaluate the expression. The answer should be in scientific notation.

$(7.0 \times 10^4)^2$

F 7.0×10^6

H 4.9×10^9

G 7.0×10^8

J 49×10^8

Hint

Use the rules of exponents and powers. Make sure the answer is in scientific notation.

33. Which one of the following shows two approximate solutions to the expression $\sqrt{36} \times \sqrt{49} + \sqrt{10}$?

A 39, 0

C 53, 35

B $-39, 39$

D 53, 0

Hint

The square root of a number can be either positive or negative.

34. Which one of the following shows some of the different ways to write b^{10}?

F $b^1 \times b^8, (b^2)^5, b^8 \times b^2$

G $b^5 \times b^5, \dfrac{1}{(b^2)^5}, b^8 \times b^1 \times b$

H $b^5 \times b^5, (b^2)^5, b^8 \times b^2$

J $b^5 \times b^5, (b^2)^{-5}, b^8 \times b^{-2}$

Hint

To raise to a power, multiply the exponents. To multiply, add exponents. Negative exponents mean the reciprocal of the number.

35. The sum of the digits of a two-digit number is 11. When the digits are reversed, the number is increased by 9. What is the number?

A 47

C 65

B 56

D 83

Hint

Write two equations, one for the sum of the digits and the other for when the digits are reversed. Remember that there is a tens place and a ones place.

36. Solve the equation.

$\dfrac{x-1}{2} = \dfrac{3}{4}$

F $x = 1.5$

H $x = 2.5$

G $x = 2$

J $x = 3.5$

Hint

Cross-multiply and solve for x by combining like terms.

37. Solve the system of equations.

$$2y = 6x + 4$$
$$y = 3x + 2$$

Which one of the following is true of the solution?

A The solution is one ordered pair.

B The solution is two ordered pairs.

C There is no solution.

D The solution is an infinite number of ordered pairs.

Hint

Use the slope-intercept form for the first equation and compare the slopes of the equations.

38. Look at the graph below.

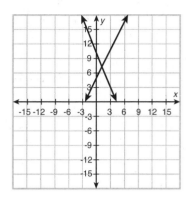

Which ordered pair represents the solution to the system of equations?

F (0, 10) **H** (0, 5)

G $(\frac{5}{4}, \frac{15}{2})$ **J** Not Here

Hint

The solution is the intersection of the two lines.

39. Look at the graph below.

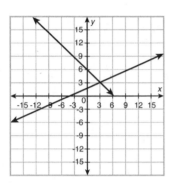

Which ordered pair represents the solution to the system of equations?

A   (0, 6)   C   (6, 0)

B   (3, 3)   D   (−4, 0)

Hint

The solution is the intersection of the two lines.

40. Look at the graph below.

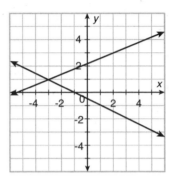

Which one of the following is true about the graph?

F   The equations represented by the graph have only one solution.

G   The equations represented by the graph have no solution.

H   The equations represented by the graph have an infinite number of solutions.

J   The equations represented by the graph have exactly two solutions.

Hint

The number of points on both pairs of lines represents the number of solutions.

41. What is the reciprocal of $\frac{1}{10^{-2}}$?

    A $\frac{1}{0.01}$ **C**   1.01

    B   0.01   **D** $\frac{1}{100}$

Hint

     $n^{-1} = \frac{1}{n}$

42. Solve the equation.

     $7(x + 3) + 3^3 - 2x = 4(x + 5) + \frac{1}{2}(2x) + 28$

How many solutions are there for this equation?

    F   0   **H**   2

    G   1   **J**   infinite number

Hint

Is the solution to the equation true or not true?

43. Look at this system of equations.

     $y = 3x + 2$
     $y = 5x - 10$

Which one of the following graphs shows the solution to the system of equations?

A

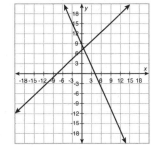

B

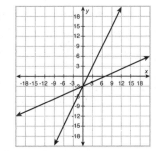

C

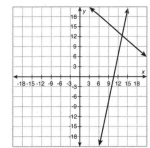

D

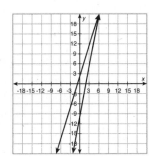

Hint

Substitute the coordinate of each pair of intersecting lines in the two equations.

Expressions and Equations
Higher Scores on Math, Grade 8

44. Solve the equation.

$$4x + 3 - 2x = 17 + 2x - 7$$

How many solutions are there for this equation?

F 0

H 2

G 1

J infinite number

Hint

Solve for x using the properties of equations.

45. Juliana is covering her square table with square tiles. If the area of the tabletop is 64 square inches, how many square tiles will be on each side and how many will she need for the edges?

A 6 on each side and 48 for the edges

B 8 on each side and 28 for the edges

C 8 on each side and 32 for the edges

D Not Here

Hint

Find the lengths of the sides. If she places tiles on the edge, what happens at the corners?

46. Joe sells CDs and DVDs. The CDs sell for $7.95 each, and the DVDs sell for $14.95 each. Last month he sold a total of 25 and made $303.75. How many of each type did he sell?

F 15 CDs and 10 DVDs

G 10 CDs and 15 DVDs

H 12 CDs and 13 DVDs

J 20 CDs and 5 DVDs

Hint

Write one equation for the number of items and one for the money.

47. Solve the system of equations.

$$y = 2x + 1$$
$$y = 4x - 1$$

Which ordered pair is the solution?

A (3, 1)

C (−3, 1)

B (1, 3)

D (1, −3)

Hint

Substitute in the equations the values of x and y in each of the ordered pairs.

48. Solve the system of equations.

$$y = 2x + 10$$
$$4y = 8x + 16$$

Which one of the following is true of the solution?

F The solution is one ordered pair.

G The solution is an infinite number of ordered pairs.

H There is no solution.

J The solution is two ordered pairs.

Hint

Divide both sides of the second equation by 4 and compare the slopes.

49. Which one of the following is the equation of the line with a slope of $\frac{7}{8}$ and a y-intercept of -4?

A $y = \frac{7}{8}x - 4$

B $y = \frac{7}{8}x + 4$

C $y = \frac{8}{7}x - 4$

D $y = \frac{8}{7}x + 4$

Hint

Substitute the slope and y intercept into the equation $y = mx + b$.

50. Solve the equation.

$$2x(x + 3) = 2x^2 + 15$$

Which one of the following is true of the solution?

F $x = 2$

G $x = 2\frac{1}{2}$

H There is no solution.

J There are an infinite number of solutions.

Hint

Use the distributive property to clear the parentheses. Then combine like terms.

Expressions and Equations

Independent Practice

DIRECTIONS: Read each question and choose the best answer. Use the answer sheet provided at the end of the workbook to record your answers. If the correct answer is not available, mark the letter for "Not Here."

51. What are the four possible solutions for the expression $\sqrt{49} + \sqrt{100} - 25$?

    A $8, -42, 28, -22$

    B $-8, -42, -28, -22$

    C $-8, 42, -28, 22$

    D $8, 42, 28, 22$

52. The mass of the moon is 7.36×10^{22} kilograms. The mass of Mars is 6.24×10^{23} kilograms. How many times greater is the mass of Mars than that of the moon?

    F 8.47 **H** 0.0117

    G 8.47×10^{2} **J** 1.17×10^{2}

53. Solve the equation.

$$4^{3}x + 56 = 8^{2}x - 36 + 9^{2}$$

How many solutions does the equation have?

    A   no solutions

    B   one solution

    C   exactly two solutions

    D   an infinite number of solutions

54. Solve the equation.

$$6.4(x + 4) + 4^{2} + 3.4(x - 3) = 54$$

    F $x = 0.22$ **H** $x = 2.3$

    G $x = 0.73$ **J** $x = 7.53$

55. Two coordinates of line A are $(0, 15)$ and $(10, 0)$. Two coordinates of line B are $(0, \frac{7}{3})$ and $(2\frac{1}{3}, 0)$. What is true about lines A and B?

    A   They intersect in one point.

    B   They intersect in two points.

    C   They are parallel.

    D   They are the same line.

56. Solve the equation.

$$3x + 25 - 2x = 5^{2} + x$$

Which one of the following describes the solution?

    F   no solution   **H** $x = 7$

    G   all real numbers   **J** $x = 5$ and $x = 4$

57. Tanisha bought peaches and pears. She bought a total of 15 items and spent a total of $6.30. The peaches cost $0.50 each, and the pears cost $0.35 each. How many of each did she buy?

    A   7 peaches and 8 pears

    B   8 peaches and 7 pears

    C   9 peaches and 4 pears

    D   3 peaches and 10 pears

58. What is $(4.3 \times 10^{3})^{4}$ in scientific notation?

    F 3.4188×10^{14} **H** 17.2×10^{3}

    G 17.2×10^{12} **J** 3.4188×10^{7}

59. Evaluate the expression for the given value of each variable.

$5a^2 + 4b^3$ when $a = 3$ and $b = 4$

A 31   **C** 188

B 109   **D** 301

60. When an equation is solved for x, the result is $7 = 21$. How many solutions are there for the equation?

F 0   **H** 2

G 1   **J** infinite number

61. Look at this system of equations.

$y = 4x + 2$
$3y = 9x + 18$

Which one of the following CANNOT be used to help solve the system of equations?

A $3(4x + 2) = 9x + 18$

B $12x + 6 = 9x + 18$

C $12x + 9x = 18 + 6$

D $12x - 9x = 18 - 6$

62. Look at the graph below.

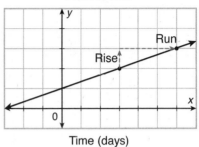

Plant Growth

Time (days)

What are the rate of change and the slope shown by the graph?

F rate of change: 3, slope: $\frac{1}{3}$

G rate of change: $\frac{1}{3}$, slope: 3

H rate of change: 3, slope: 3

J rate of change: $\frac{1}{3}$, slope: $\frac{1}{3}$

63. What is the standard notation for 5.7832×10^{-3}?

A 0.00057832   **C** 5,783.2

B 0.0057832   **D** 578,320

64. What is the slope of the equation $6x + 3y = 36$?

F -6 **H** 2

G -2 **J** 6

6. Solve the system of equations.

$y = 5x - 8$
$5y = 25x + 40$

Which one of the following is true of the solution?

A   The solution is one ordered pair.

B   The solution is two ordered pairs.

C   There is no solution.

D   The solution is an infinite number of ordered pairs.

66. Which of the data tables below gives some solutions for the equation: $y = x - (-4x + 7)$?

F

x	1	3	7	9
y	−9	9	45	62

G

x	5	13	19	20
y	27	99	153	160

H

x	2	6	8	1
y	0	36	54	71

J

x	2	4	6	8
y	4	22	40	58

67. When an equation is solved for x, the result is $10 = 10$. How many solutions are there for this equation?

A   no solutions

B   one solution

C   two solutions

D   an infinite number of solutions

68. Evaluate the expression.

$5^2 + (3 + 2)^3 \times (3 + 2)^{-3}$

F   5   **H**   235

G   26   **J**   275

69. Look at the graph and the table below. They show how much two people are paid for doing the same work.

Person A

Number of Lawns	1	2	3	4
Earnings ($)	15	30	45	60

Person B

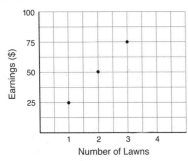

What are the rates shown in the table and graph, and who earns more?

A

Person A	Person B	Who Earns More?
25	15	Person B

B

Person A	Person B	Who Earns More?
15	25	Person B

C

Person A	Person B	Who Earns More?
25	15	Person A

D

Person A	Person B	Who Earns More?
15	25	Person A

70. What is 1,246,912 written in scientific notation?

F 1.246912×10^6 **H** 12.46912×10^5

G 1.246912×10^{-6} **J** 0.1246912×10^7

71. If the slope of a line is $\frac{7}{8}$, what two points are on the line?

A $(1, \frac{7}{8})$ and $(32, 27)$

B $(2, 1\frac{1}{2})$ and $(4, 3\frac{1}{7})$

C $(24, 21)$ and $(24, 20)$

D $(8, 7)$ and $(16, 14)$

72. Solve the system of equations.

$3y = 6x + 3$
$5y = 2x + 5$

Which ordered pair falls on the graph of the two lines?

F $(0, -\frac{1}{2})$ **H** $(0, -1)$

G $(0, 1)$ **J** $(-2\frac{1}{2}, 0)$

73. Solve the equation.

$26 - 2(2x + 2) = 4(x + 2) + 6x$

A $x = -14$ **C** $x = 5$

B $x = -5$ **D** $x = 1$

74. Four times a number is equal to 7 plus 8 times the number. What is the number?

F $-1\frac{3}{4}$ **H** $\frac{7}{12}$

G $-\frac{7}{12}$ **J** $1\frac{3}{4}$

75. Solve the system of equations.

$$2y = 4x + 20$$
$$12y = 24x + 48$$

Which one of the following is true for the solution?

A   The solution is two ordered pairs.

B   The solution is one ordered pair.

C   There is no solution.

D   The solution is an infinite number of ordered pairs.

76. What is 0.00007629 written in scientific notation?

F 7.629×10^{-5} **H** 76.29×10^{-6}

G 7.629×10^{5} **J**   Not Here

77. Solve the system of equations.

$$y = 3x + 2$$
$$y = 5x - 10$$

Which ordered pair is the solution?

A   (20, 6)   **C**   (6, 20)

B   (0, 2)   **D**   (0, −10)

78. The Indian Ocean covers approximately 6.8556×10^7 square kilometers. The Atlantic Ocean covers approximately 1.06448504×10^8 square kilometers. The Atlantic Ocean is how many times larger than the Indian Ocean?

F 1.55272×10^{2} **H** 1.55272×10^{3}

G 1.55272×10^{0} **J** 1.55272×10^{-2}

79. Which does NOT have the same solution as $7(3 - x) - 4(2 - 3x) = 30$?

A $5x + 13 = 30$

B $5x = 17$

C $-5x - 13 = -30$

D $5x = -17$

80. Solve this system of equations.

$$2y = 6x + 4$$
$$y = 3x + 2$$

Which one of the following is true about the solution?

F   The solution is one ordered pair.

G   The solution is two ordered pairs.

H   The solution is all rational numbers.

J   There is no solution.

81. Which of the equations below has exactly one solution?

A $3(5 - x) = 2x - 5x + 14$

B $4x + 3(x - 5) = 3x + 26$

C $2^2 + 3x - 5 = 3x + 2^2 - 5$

D $x(4 + 2) + 3^2 = 2x + 4x + 2^3$

82. Solve the equation.

$$2x + 10 = \frac{3}{5}x + \frac{108}{5}$$

How many solutions does this equation have?

F   0   **H**   2

G   1   **J**   infinite number

83. Solve this equation.

$$12(x - 5) + 3x = 5(3x + 4) - 2(x + 5)$$

How many solutions does this equation have?

A 0

C 2

B 1

D infinite number

84. The graph below shows two similar triangles, ABC and $A'B'C'$.

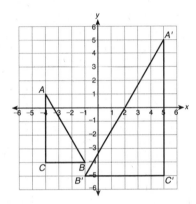

What is the relationship between AB and $A'B'$?

F The slopes are equal.

G The slopes are both positive and unequal.

H The slopes are equal, but one is positive and one is negative.

J Both slopes are negative.

85. Which one of the following shows these numbers in order from least to greatest?

$$\sqrt[3]{125}, \sqrt{100}, \sqrt[3]{343}, \sqrt{81}$$

A $\sqrt{81}, \sqrt{100}, \sqrt[3]{125}, \sqrt[3]{343}$

B $\sqrt[3]{343}, \sqrt{81}, \sqrt{100}, \sqrt[3]{125}$

C $\sqrt[3]{125}, \sqrt{81}, \sqrt{100}, \sqrt[3]{343}$

D $\sqrt[3]{125}, \sqrt[3]{343}, \sqrt{81}, \sqrt{100}$

86. Tyrone buys books for the library. He pays $6.95 for children's books and $10.27 for adult fiction books. He spent $100.00 on his last order of 12 books. How many of each kind of book did he order?

F 7 books for children and 5 for adults

G 5 books for children and 7 for adults

H an equal number of each book

J 8 books for adults and 4 for children

87. The graph below shows time and distance for a bicycle rider.

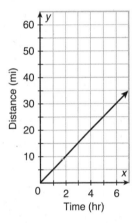

What are the rate of change and slope of the line?

A rate of change: $\frac{1}{5}$, slope: $\frac{1}{5}$

B rate of change: 5, slope: $\frac{1}{5}$

C rate of change: $\frac{1}{5}$, slope: 5

D rate of change: 5, slope: 5

88. Solve the expression.

$$4.7327 \times 10^{10} \times 6 \times 10^8$$

F 1.07327×10^{19}

H 2.83962×10^{19}

G 10.7327×10^{18}

J 28.3962×10^{18}

89. A number times 2 plus a second number equals 125. The first number is 3 times the second number. What are the numbers?

A   31.25 and 46.87   **C**   53.86 and 17.57

B   53.57 and 17.86   **D**   31.87 and 46.25

90. A cube has a volume of 1,728 cm³. What is the length of each side?

F   12 cm   **H**   288 cm

G   144 cm   **J**   576 cm

1. Look at the graph below.

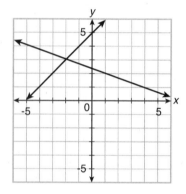

Which ordered pair is the solution to the system of equations represented by the lines on the graph?

A $(-5, 0)$ **C** $(4, 1)$

B $(0, 5)$ **D** $(-2, 3)$

92. Look at the graph below.

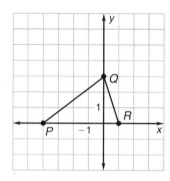

What is the slope of PQ?

F $-\dfrac{3}{4}$ **H** $\dfrac{4}{3}$

G $\dfrac{3}{4}$ **J** -3

93. Which one of the following shows some of the different ways to write 25^{-4}?

A $25^5 \times 25^{-9}, (25^2)^{-2}, 25^{-10} \times 25^6$

B $\dfrac{1}{25}^{-5} \times 25^{-9}, (25^2)^2, 25^{-10} \times 25^{-6}$

C $25^5 \times 25^9, (25^2)^2, 25^{-10} \times 25^6$

D $29^{-9} \times 25^{-5}, \left(\dfrac{1}{25}\right)^2, 25^{10} \times 25^{-6}$

94. Which one of the following is the equation of the line with a slope of 3 and a y-intercept of -15?

F $y = 3x + 45$ **H** $y = \dfrac{1}{3}x + -15$

G $3y = 9x - 45$ **J** $2y = 7x - 30$

Expressions and Equations
Higher Scores on Math, Grade 8

95. Which graph below shows a system of equations with an infinite number of solutions?

A
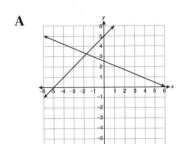

B
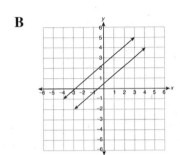

C
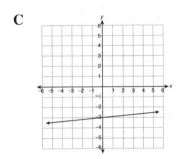

D
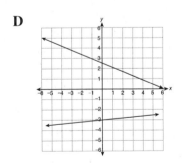

96. Which one of the following is the solution to the expression $5.7 \times 10^2 + 3.5 \times 10^6$?

F 9.2×10^6

G 9.2×10^8

H 3.50057×10^7

J 3.50057×10^6

97. Which of the data tables below gives some of the solutions to the equation $y = 3x + 25$?

A

x	0.5	1	1.5	2	2.5
y	26.5	28	29.5	30	32

B

x	3	7	12	15	19
y	34	46	61	70	80

C

x	7	9	11	13	15
y	46	52	58	64	70

D

x	6	7	8	9	10
y	40	47	49	52	53

98. The chart and table below are tracking how long it takes to fill two pools.

Pool A

Time	8	10	12	14
Gallons	56	70	84	98

Pool B

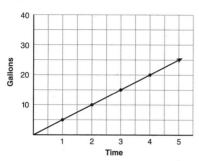

Which one of the following best describes what is shown by the data?

F

	Pool A	Pool B
Slope	5	7
Speed		Fills Faster

G

	Pool A	Pool B
Slope	7	5
Speed	Fills Faster	

H

	Pool A	Pool B
Slope	7	5
Speed		Fills Faster

J

	Pool A	Pool B
Slope	5	7
Speed	Fills Faster	

99. Two similar triangles ABC and $A'B'C'$ are drawn on a coordinate plane. Which one of the following is NOT true?

A The corresponding sides of the triangles have the same slope, although one may be positive and one may be negative.

B If the triangles are right triangles, the hypotenuse of one has a different slope than the hypotenuse of the other.

C The respective angles in the two triangles are equal.

D The lengths of their corresponding sides are proportional.

100. Look at the graph below.

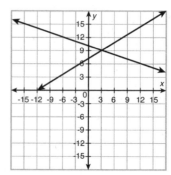

Which ordered pair is the solution to the system of equations represented by the graph?

F $(-12, 0)$

G $(3, 9)$

H $(-6, 12)$

J $(12, 6)$

101. What is the reciprocal of $-\dfrac{5}{6^{-3}}$?

A $-\dfrac{1}{1,080}$ **C** $-1,080$

B $43\dfrac{1}{5}$ **D** Not Here

102. Solve the equation.

$$30(50 - 10x) - 20(50 + 50x) = 30(10x + 1)$$

F -1.175

H 0.331

G 0.293

J 0.361

103. Evaluate the expression.

$$(5.3 \times 10^{-3})^4$$

A 7.89048×10^{-10}

C 7.89048×10^1

B 7.89048×10^{10}

D 7.89048×10^{-7}

104. Graph the system of equations.

$$y = x + 4$$
$$y = -x$$

F
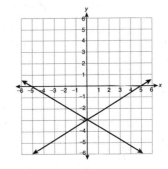

G
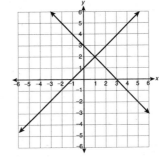

H
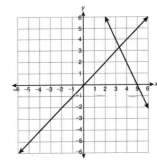

J
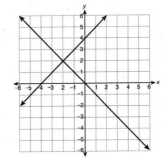

Expressions and Equations
Higher Scores on Math, Grade 8

Functions

Modeled Instruction

DIRECTIONS: Read each question and choose the best answer. Use the answer sheet provided at the end of the workbook to record your answers. If the correct answer is not available, mark the letter for "Not Here."

1. Which one of the following sets of ordered pairs is NOT a function?

   A (1, 4), (2, 8), (3, 12), (4, 16)

   B (3, 195), (5, 325), (9, 585), (10, 650)

   C (2, 5), (5, 11), (5, 17), (10, 21)

   D (2, 15), (3, 22.5), (5, 37.5), (8, 60)

   Hint

   A function can have only one output for each input.

Name _____ Date _____

2. The equations below define two different functions.

Function 1: $y = 2x + 5$ Function 2: $y = 4x - 3$

What is the rate of change for each function, and which function has the greater rate of change?

F

Function 1	Function 2	Greater Rate of Change
2	4	Function 2

G

Function 1	Function 2	Greater Rate of Change
2	4	Function 1

H

Function 1	Function 2	Greater Rate of Change
4	2	Function 1

J

Function 1	Function 2	Greater Rate of Change
4	2	Function 2

Hint

Look at the slope of each line.

3. Which of the tables below displays a nonlinear function?

A

1	3	5	7	9
6	8	10	12	14

B

1	2	3	4	5
1	8	27	64	125

C

1	2	3	4	5
−4	−3	−2	−1	0

D

7	8	9	10	11
17	19	21	23	25

Hint

Look for a relationship in each table.

4. Jody gets paid $200.00 per week, plus $10.00 for every oil change she completes. Which equation shows the relationship between the number of oil changes and her total weekly income?

F $2y = 400 + 20x$

G $y = 200 - 10x$

H $\frac{1}{2}y = 20x + 2{,}000$

J $y = 200 + 5x$

Hint

The total of her salary added to what she makes per oil change is her weekly income.

5. Which one of the following graphs represents a situation in which the amount spent on outdoor maintenance was low during the first quarter of the year, gradually rose during the second quarter, stabilized during the third quarter, and dropped during the fourth quarter?

A

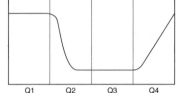

B

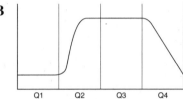

C

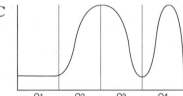

D

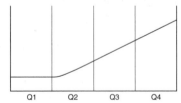

Hint

Rising on the graph is up, and falling is down.

Name _____ Date _____

6. Look at this table.

x	1	2	5	6	8
y	7	9	a	17	b

What values of a and b will make this a function?

F $a = 10$ and $b = 23$

G $a = 11$ and $b = 19$

H $a = 15$ and $b = 21$

J $a = 13$ and $b = 21$

Hint

Find the relationship between all the values of x and y.

7. Look at the tables below.

Distance (mi)	$\frac{1}{2}$	1	$1\frac{1}{2}$	2	4
Time (hr)	$\frac{1}{4}$	$\frac{1}{2}$	$\frac{3}{4}$	1	2

Tyler

Distance (mi)	$\frac{1}{3}$	$\frac{2}{3}$	1	2	4
Time (hr)	$\frac{1}{6}$	$\frac{1}{3}$	$\frac{1}{2}$	1	2

Tonia

Which statement is true for these data?

A Tyler is walking faster than Tonia.

B Tonia is walking faster than Tyler.

C Tonia and Tyler are walking at the same speed.

D Tyler is walking twice as fast as Tonia.

Hint

Compare the rates of change.

8. The graph below represents a function.

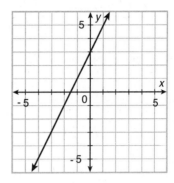

Which equation describes the function?

F $y = 3x + 2$ **H** $y = 2x + 3$

G $y = 2x - 3$ **J** $y = -3x + 2$

Hint

Find the slope and y-intercept of the line.

9. Oscar and Mary both work in a garage. Mary's weekly income can be described by the function $y = 25x + 100$, where y is her salary and x is the amount she receives for every tire she changes. Oscar's weekly income is shown in the table below.

Tires Changed	5	6	7	8
Income ($)	235	260	285	310

Who is paid more?

A Oscar is paid more than Mary.

B Mary is paid more than Oscar.

C They are paid the same amount.

D There is not enough information given to answer the question.

Hint

Write a function for the data in the table or create a data table for the function given.

10. Look at the graph.

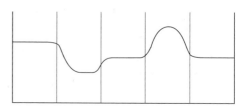

Which one of the following is the best interpretation of the graph?

F Mr. Lauve graphs the test scores of his students from September to June. He finds the scores continue to rise until December. In January, they fall. They rise again in February and steadily increase until June.

G Mrs. Reyna graphs attendance at school activities. Attendance is high during the football season but gradually declines for the rest of the school year.

H Jon works for a company that has seasonal sales selling snowplows in the winter and lawn maintenance equipment in the summer.

J Not Here

Hint

Sketch a graph of each answer choice and compare each to the one given.

Functions

Independent Practice

DIRECTIONS: Read each question and choose the best answer. Use the answer sheet provided at the end of the workbook to record your answers. If the correct answer is not available, mark the letter for "Not Here."

11. Which of the graphs below represents a situation in which the number of items shipped is high in the first quarter, low in the second and third quarters, and rises during the fourth quarter?

A
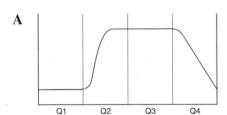

B
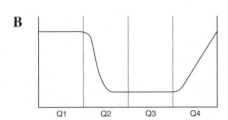

C
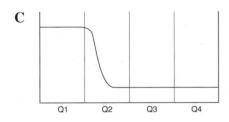

D
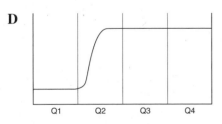

12. Look at the graph below.

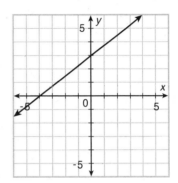

Which one of the following describes this function?

F $y = \frac{3}{4} + 3$ **H** $y = -\frac{3}{4}x + 3$

G $y = \frac{4}{3}x + 3$ **J** $y = -\frac{4}{3}x - 3$

13. Laurel and Eli both work in sales. Laurel's weekly income can be described by the function $y = 10x + 100$, where y is her salary and x is the number of items she sells. Eli's weekly income is shown in the table below.

Number of Items	3	5	7
Total Income ($)	180	200	220

Who has the higher salary?

A Laurel's salary is higher than Eli's.

B Eli's salary is higher than Laurel's.

C Their salaries are the same.

D Not enough information is given to solve the problem.

14. Isaac gets paid $400.00 per week, plus a commission of $10.00 for every item he sells. Which equation shows the relationship between the number of items he sells and his weekly income?

F $y = 200 - 5x$

G $\frac{1}{2}y = 200 + 5x$

H $y = 200 + 5x$

J $y = 400 - 10x$

15. Which table below describes a nonlinear function?

A

2	3	4	6
8	27	64	216

B

12	15	17	19
12	15	17	90

C

1	3	5	7
8	14	20	26

D

2	4	6	8
-4	0	4	8

16. The equations below define two different functions.

Function 1: $y = \frac{5}{2}x + 12$

Function 2: $y = -\frac{1}{2}x$

What is the rate of change of each function, and which function has the greater rate of change?

F

Function 1	Function 2	Greater Rate of Change
$\frac{5}{2}$	$-\frac{1}{2}$	Function 2

G

Function 1	Function 2	Greater Rate of Change
$\frac{5}{2}$	$-\frac{1}{2}$	Function 1

H

Function 1	Function 2	Greater Rate of Change
$-\frac{1}{2}$	$\frac{5}{2}$	Function 2

J

Function 1	Function 2	Greater Rate of Change
$-\frac{1}{2}$	$\frac{5}{2}$	Function 1

17. Which table below defines a nonlinear function?

A

2	6	8	10
10	22	28	34

B

1	3	9	12
9	27	81	108

C

2	7	9	20
4	49	81	400

D

6	8	9	12
25	31	34	43

18. The equation below shows how the total distance Elia walks to school depends on the number of trips she makes. The total distance she has walked is represented by d. The number of trips is represented by t.

$$d = 2t + 25$$

How many trips will Elia have to make to walk a total of 49 miles?

F 12   **H** 50

G 25   **J** 75

19. Which one of the following ordered pairs does NOT define a function?

A (3, 4), (5, 6), (7, 8), (9, 10)

B (10, 5), (8, 4), (7, 3.5), (5, 2.5)

C (5, 7), (6, 9), (7, 10), (7, 12)

D ($\frac{1}{2}$, 1), (1, 1.5), (1.5, 2), (2, 2.5)

20. The graph below shows two linear functions.

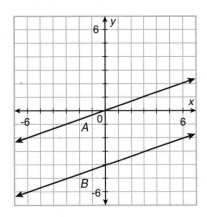

Which one of the following is true?

F Line A has a greater rate of change than line B.

G Line B has a greater rate of change than line A.

H The rates of change are the same and equal to $\frac{1}{2}$.

J The rates of change are the same and equal to $\frac{1}{3}$.

21. Look at the tables below.

Distance (mi)	1	2	3	4	5	6
Time (hr)	0.5	1	1.5	2	2.5	3

Violet

Distance (mi)	1	2	3	4	5	6
Time (hr)	0.75	1.5	2.24	3	3.75	4.5

Rose

Which statement is true for these data?

A Violet is faster than Rose.

B Rose is faster than Violet.

C Violet and Rose walk at the same rate.

D Rose walks twice as fast as Violet.

Name _____   Date _____

22. Look at the table below.

−2	−1	0	3	6	9
4	2	0	a	−12	b

What values for a and b will make this a function?

F $a = 1$ and $b = \dfrac{1}{2}$

G $a = -6$ and $b = -18$

H $a = 2$ and $b = 1$

J $a = 4$ and $b = 2$

23. Look at the graph.

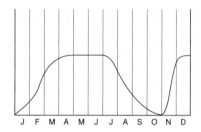

Which one of the following is the best interpretation of the graph?

A Mr. Chang's water bill tracks his water usage. He uses a lot of water in January, February, and March. He uses almost no water in April through September and then gradually increases water usage to the end of the year.

B Mrs. Frank's water bill shows a gradual increase in January and February, a high usage from April to June, and a decline until November. Then it shows an increase to the end of the year.

C Mr. Foster's water usage goes up and down during alternating months.

D Ms. Salazar's water usage goes down from January to November and rises through November and December.

24. The graph below represents a function.

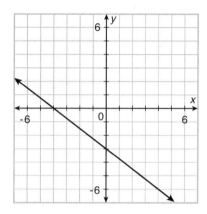

Which equation describes the function shown by the line?

F $-2y = \dfrac{3}{2}x + 6$

G $2y = \dfrac{3}{2}x - 12$

H $-4y = 3x + 12$

J $y = 3x - 3$

25. Which one of the following ordered pairs defines a function?

A $(6, 2), (6, 5), (7, 21)$

B $(12, 4), (15, 5), (15, 7)$

C $(8, 2), (7, 3), (7, 9)$

D $(5, 5.5), (8, 7.5), (10, 8)$

26. Look at the table and the graph below.

Figure A

x	1	3	5	7	9
y	3	9	15	21	27

Figure B

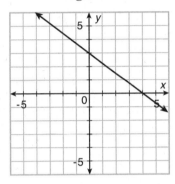

What are the rates of change for the two figures, and which is greater?

F

Figure A	Figure B	Greater Rate of Change
$\frac{1}{3}$	$-\frac{3}{4}$	Figure A

G

Figure A	Figure B	Greater Rate of Change
3	$-\frac{3}{4}$	Figure A

H

Figure A	Figure B	Greater Rate of Change
3	$\frac{3}{4}$	Figure A

J

Figure A	Figure B	Greater Rate of Change
-3	$-\frac{3}{4}$	Figure B

27. Which of the graphs shows that sales were high throughout the year except during the summer months?

A

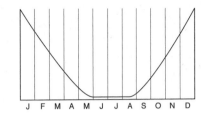

B

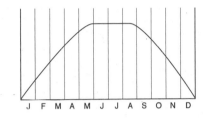

C

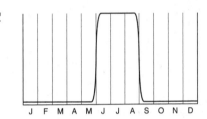

D

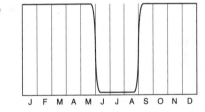

28. Look at the graph below.

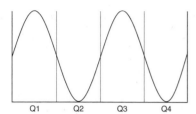

Which one of the following is the best situation for this graph?

F   Temperatures were up and down throughout the year.

G   Temperatures were consistently high throughout the year.

H   Temperatures were consistently low throughout the year.

J   Temperatures were low in the first two quarters and high in the last two quarters.

---

29. The equations below define two different functions.

Function 1: $y = -\dfrac{7}{5}x + 3$ Function 2: $y = -2x - \dfrac{7}{2}$

What are the rates of change for the two functions, and which has the greater rate of change?

A

Function 1	Function 2	Greater Rate of Change
$\dfrac{7}{5}$	2	Function 1

C

Function 1	Function 2	Greater Rate of Change
$-\dfrac{7}{5}$	-2	Function 2

B

Function 1	Function 2	Greater Rate of Change
$\dfrac{7}{5}$	2	Function 2

D

Function 1	Function 2	Greater Rate of Change
$-\dfrac{7}{5}$	-2	Function 1

---

30. Look at the table below.

-2	0	2	3	4
a	-7	b	-8.5	-9

What values of a and b will make this a linear function?

F $a = 6$ and $b = 8$ **H** $a = -4$ and $b = -3$

G $a = -6$ and $b = -8$ **J** $a = 10$ and $b = -11$

Geometry

Modeled Instruction

DIRECTIONS: Read each question and choose the best answer. Use the answer sheet provided at the end of the workbook to record your answers. If the correct answer is not available, mark the letter for "Not Here."

1. Look at the graph below.

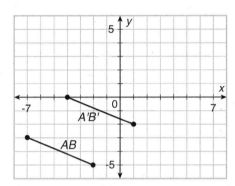

Which one of the following is true about the line segments AB and $A'B'$?

A AB is translated to $A'B'$ by going 3 over and 3 up.

B AB is not as long as $A'B'$.

C The slope of AB is not the same as the slope of $A'B'$.

D If these segments are extended to make lines, then they will eventually intersect.

Hint

The segments changed positions but remained parallel and the same length.

2. Triangle ABC is transformed into triangle $A'B'C'$ by moving it 3 down and 5 across. What is this called?

F a dilation

G a translation

H a rotation

J a reflection

Hint

Which of the choices names a transformation that just moves the figure?

3. What is the volume of a cylinder 10 feet high and 8 feet across? Use $\pi = 3.14$.

A 251.2 ft³

B 502.4 ft³

C 1,256 ft³

D 128,614.4 ft³

Hint

Use the formula $V = Bh$ or $V = \pi r^2 h$.

4. Look at the graph below.

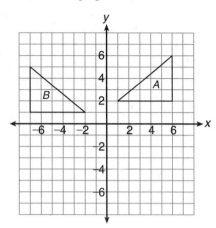

Which series of transformations took place for figure A to become figure B?

F dilation and reflection

G translation and dilation

H dilation and rotation

J reflection and translation

Hint

Compare the position and size of figure B to figure A.

5. Figure *ABCDE* is dilated with a scale factor of $\frac{1}{2}$ to form figure *A'B'C'D'E'*. Which one of the following is true?

A The angles in *ABCDE* are not equal to the respective angles in *A'B'C'D'E'*.

B The sides of *ABCDE* are equal to the respective sides of *A'B'C'D'E'*.

C The parallel lines in *ABCDE* are also parallel in *A'B'C'D'E'*.

D Figure *ABCDE* is congruent to *A'B'C'D'E'*.

Hint

In a dilation, angle measures and parallel lines remain unchanged. Line size does change.

6. Look at the graph below.

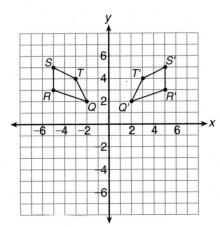

What is NOT true about the figures *QRST* and *Q'R'S'T'*?

F *Q'R'S'T'* is a dilation of *QRST*.

G *Q'R'S'T'* is a reflection of *QRST*.

H *Q'R'S'T'* is congruent to *QRST*.

J *SR* is parallel to *S'R'*.

Hint

The figures are in a different position, but they are the same size and shape.

7. Look at the graph below. *ABCD* is rotated and translated to form *A'B'C'D'*.

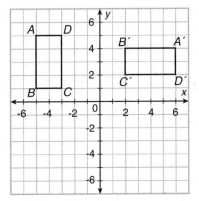

Which one of the following is NOT true?

A *AD = A'D'*

B *ABCD* is congruent to *A'B'C'D'*.

C *AB ∥ A'B'*

D *CD = C'D'*

Hint

Rotation does not affect congruency, but it does affect parallelism.

8. Which transformations changed figure A into figure B?

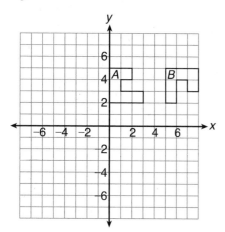

F translation followed by rotation

G reflection followed by translation

H translation followed by dilation

J rotation followed by reflection

Hint

Determine the difference between the two figures and see how they got that way.

9. A cylinder and a cone have the same height and the same radius of their base. If their height is 20 cm and the diameter of their base is 10 cm, which one of the following statements is true about the volume of the two figures? Use $\pi = 3.14$.

A The two volumes are equal.

B The volume of the cylinder is three times larger than the volume of the cone.

C The volume of the cone is three times larger than the volume of the cylinder.

D Not Here

Hint

Find the volume of each and compare.

10. There are two lines on a coordinate graph. *AB* has coordinates $(-4, 0)$ and $(-8, 2)$. $A'B'$ has coordinates $(-1, 2)$ and $(-5, 4)$. Which one of the following statements is NOT true?

F The two lines are parallel.

G $A'B'$ was formed by translating *AB*.

H The two lines are perpendicular.

J The two lines do not intersect.

Hint

Draw the lines on a coordinate plane.

11. What is the length of the side of a cone if the radius of the base is 12 and the height is 20?

A $\sqrt{32}$ **C** $\sqrt{544}$

B $\sqrt{250}$ **D** $\sqrt{976}$

Hint

Use the formula $a^2 + b^2 = c^2$.

12. How many cubic centimeters are there in a cone that has a base 2 centimeters across and a height of 10 centimeters? Use $\pi = 3.14$.

F 41.86 cm³   **H** 20.93 cm³

G 10.4 cm³   **J** 10.46 cm³

Hint

Use the formula $V = \frac{1}{3} Bh$.

13. Quadrilateral $ABCD$ is rotated 90° in a clockwise direction forming quadrilateral $A'B'C'D'$. Which one of the following is true?

A The sum of the angles of $A'B'C'D'$ equals 180°.

B The sum of the angles of $A'B'C'D'$ equals 360°.

C The sum of the angles of $A'B'C'D'$ is greater than the sum of the angles of $ABCD$.

D The sum of the angles of $A'B'C'D'$ is less than the sum of the angles of $ABCD$.

Hint

The sum of the angles in a quadrilateral is 360°.

14. Look at the graph below.

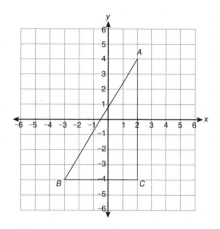

What is the length of side AB to the nearest tenth?

F 3.6   **H** 9.4

G 6.2   **J** 13

Hint

Use the formula $a^2 + b^2 = c^2$.

15. Mr. Hernández needs to pack cylinders into a carton that is 12 in. × 12 in. × 12 in. What is the maximum number of cylinders he can get into the carton if the cylinders are 8 in. high and 10 in. across? Use $\pi = 3.14$.

A 1   **C** 2.75

B 2   **D** 3

Hint

Use the formulas $V_{cube} = s^3$ and $V_{cylinder} = Bh$.

Name _____ Date _____

16. Look at the graph below.

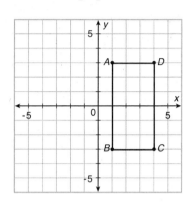

What is the length of diagonal *BD* of the rectangle to the nearest tenth?

F   3   **H**   6.7

G   5.1   **J**   9

Hint

Use the formula $a^2 + b^2 = c^2$.

17. Look at the figure below. Triangle *ABC* is an isosceles triangle.

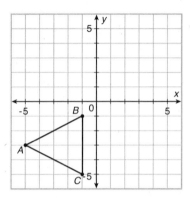

What is the length of one of the two equal sides of the triangle to the nearest tenth?

A   2.6   **C**   7

B   4.5   **D**   Not Here

Hint

Use the formula $a^2 + b^2 = c^2$.

18. Rectangle *ABCD* is rotated 90° clockwise to form *A′B′C′D′*. Which one of the following statements is NOT true?

F   Rectangles *ABCD* and *A′B′C′D′* are congruent.

G   Opposite sides of *A′B′C′D′* are parallel.

H   Corresponding sides of *ABCD* and *A′B′C′D′* are parallel.

J   The ratio of *ABCD* to *A′B′C′D′* is 1.

Hint

Rotating a figure changes the parallel sides.

Geometry
Higher Scores on Math, Grade 8

19. A cube is 4 feet on each side. How long is the diagonal of the cube to the nearest tenth?

A 5.7 ft   **C** 8.6 ft

B 6.9 ft   **D** 12 ft

Hint

Use the formula $a^2 + b^2 = c^2$.

20. The hypotenuse of a right triangle is 15 in. One of the sides is 5 in. What is the length of the third side?

F $\sqrt{20}$

G 10

H $\sqrt{200}$

J $\sqrt{250}$

Hint

Use the formula $a^2 + b^2 = c^2$.

21. What is the volume of a sphere with a diameter of 24 cm? Use $\pi = 3.14$.

A 150.72 cm³

B 602.88 cm³

C 7,234.56 cm³

D 57,876.48 cm³

Hint

Use the formula $V = \dfrac{4}{3}\pi r^3$.

22. Figure *ABCDE* is translated 3 down and 3 to the left, forming figure $A'B'C'D'E'$. Which one of the following is a true statement?

F $\angle A = \angle A'$, $\angle B = \angle B'$, $\angle C = \angle C'$, $\angle D = \angle D'$, and $\angle E = \angle E'$.

G The sum of angles *A*, *C*, and *E* is greater than the sum of angles A', C', and E'.

H The sum of angles *A*, *B*, *C*, *D*, and *E* is greater than the sum of angles of A', B', C', D', and E'.

J The sum of the angles of $A'B'C'D'E'$ is less than the sum of the angles of *ABCDE*.

Hint

The angles do not change when a figure is translated.

23. The vertices of $\triangle ABC$ are $A = (-6, -2)$, $B = (-8, -6)$, and $C = (-1, -6)$. If the triangle is translated 2 to the right and 2 up, what are the coordinates of $\triangle A'B'C'$?

A $A = (0, -4)$, $B = (-4, -6)$, $C = (-4, 1)$

B $A = (-5, -1)$, $B = (-7, -5)$, $C = (0, -5)$

C $A = (6, -2)$, $B = (1, -6)$, $C = (8, -6)$

D $A = (-4, 0)$, $B = (-6, -4)$, $C = (1, -4)$

Hint

Add 2 to the coordinates of each point.

24. Look at the figure below. $AB \parallel CD$.

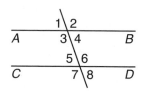

Which one of the following lists pairs of alternate interior angles?

F ∠5 and ∠4

G ∠7 and ∠2

H ∠6 and ∠4

J ∠4 and ∠1

Hint

Alternate interior angles are between the two parallel lines and on opposite sides of the transversal.

25. Look at the graph.

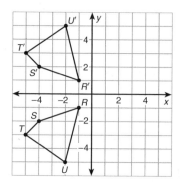

Which one of the following explains why *RSTU* is congruent to $R'S'T'U'$?

A $R'S'T'U'$ is a reflection of *RSTU* over the *x*-axis.

B $R'S'T'U'$ is a rotation of *RSTU* 90° clockwise.

C $R'S'T'U'$ is a translation of *RSTU* 3 over and 1 up.

D $R'S'T'U'$ is a dilation of *RSTU*.

Hint

Dilations change size; translations, reflections, and rotations change the coordinates of the vertices but not the size of the lines.

26. Look at the graph below.

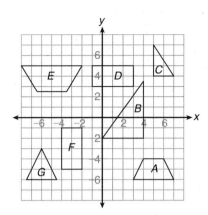

Which statement is NOT true?

F Figure B is similar to figure C as the result of a dilation, reflection, and translation.

G Figure A is similar to figure E as the result of a reflection, dilation, and translation.

H Figure F is similar to figure D as the result of a translation and rotation.

J Figure G is similar to figure C as the result of a translation and reflection.

Hint

Similar figures have the same shape but not the same size.

27. Look at the graph below.

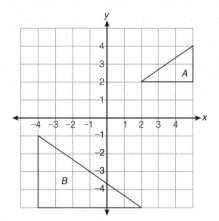

What transformations were used to create figure B from figure A?

A dilation scale factor $\frac{1}{2}$, reflection, translation

B dilation scale factor 2, reflection, translation

C reflection, rotation, translation

D dilation scale factor 2, rotation, translation

Hint

List the changes made to figure A to get to figure B.

28. Trapezoid $ABCD$ is reflected across the x-axis and then rotated 90° clockwise to form trapezoid $A'B'C'D'$. $AD \parallel BC$ and $AB \nparallel DC$. Which one of the following is NOT true?

F $AD \nparallel A'D'$ **H** $A'B' \nparallel D'C'$

G $A'D' \parallel B'C'$ **J** $A'D' \nparallel B'C'$

Hint

A reflection does not change parallelism, but a rotation does.

29. A ladder 25 feet tall is leaning against a water tank. The base of the ladder is 8 feet from the base of the tank. Approximately how far up the water tank will the ladder reach? Choose the best approximation.

A between 22 and 23 feet

B between 23 and 24 feet

C between 24 and 25 feet

D between 25 and 26 feet

Hint

Use the formula $a^2 + b^2 = c^2$.

30. If a pool table measures 4 ft by 8 ft, what is the diagonal length from the back edge of the top left pocket to the bottom right pocket? Round your answer to the nearest tenth.

F 80 ft

G 24.3 ft

H 6.9 ft

J 8.9 ft

Hint

Use the formula $a^2 + b^2 = c^2$.

31. Look at the graph below.

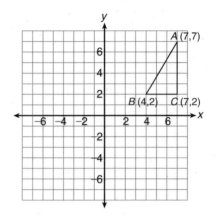

What are the coordinates of triangle $A'B'C'$ if ABC is reflected over the y-axis?

A $(-7, 7), (-4, 2), (-7, 2)$

B $(-7, 7), (2, 4), (2, -7)$

C $(-7, -7), (-4, -2), (-7, -2)$

D $(-7, -7), (-2, -4), (-2, -7)$

Hint

In a reflection over the y-axis, only the x coordinates change.

32. In the figure below, $AB \parallel CD$.

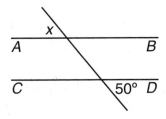

What is the measure of $\angle x$?

F 50°   **H** 90°

G 60°   **J** 150°

Hint

Vertical angles and alternate interior angles are equal.

33. What is the volume of a cylindrical bottle that is 3 inches across and 10 inches high?

A 70.65 in.³

B 282.6 in.³

C 706.5 in.³

D 1,130.4 in.³

Hint

Use the formula $V = Bh$ or $\pi r^2 h$.

34. Look at the graph below. Figure *LMN* is translated to form figure *L'M'N'* by moving *LMN* 3 to the left and 1 up.

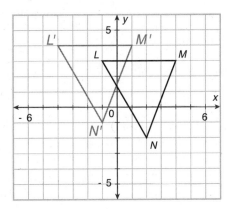

Which one of the following is a true statement?

F $LM \parallel L'M'$, $NM \parallel N'M'$, and $LN \parallel L'N'$

G $LM \nparallel L'M'$, $NM \parallel N'M'$, and $LN \parallel L'N'$

H $LM \nparallel L'M'$, $NM \nparallel N'M'$, and $LN \nparallel L'N'$

J Not Here

Hint

In a translation, lines parallel to one another remain parallel.

35. Parallelogram $A'B'C'D'$ is formed by a dilation of parallelogram $ABCD$. Which one of the following is NOT true?

  A The angles of $A'B'C'D'$ are equal to the respective angles of $ABCD$.

  B The opposite angles of $A'B'C'D'$ are equal to the respective opposite angles of $ABCD$.

  C The sum of the angles of $A'B'C'D'$ is equal to the sum of the angles of $ABCD$.

  D The sides of $A'B'C'D'$ are equal to the respective sides of $ABCD$.

Hint

In a dilation, the measures of the sides change but the measures of the angles do not.

36. Which one of the following transformations will NOT result in congruent figures?

  F dilation when the scale factor is 1

  G translation

  H reflection

  J dilation when the scale factor is not equal to 1

Hint

Congruent figures are the same shape and size.

37. Which one of the following transformations will result in figures that are similar but not congruent?

  A translation and reflection

  B dilation when the scale factor is not equal to 1

  C dilation with scale factor 1

  D rotation and reflection

Hint

All transformations except dilations result in congruent figures.

38. The vertices of triangle ABC are $A = (7, 7)$, $B = (4, 2)$, and $C = (7, 2)$. Triangle ABC is reflected across the x-axis forming triangle $A'B'C'$. Which one of the following is true?

  F $\angle A = \angle A',\ \angle B > \angle B',\ \angle C < \angle C'$

  G $\angle A + \angle B + \angle C > \angle A' + \angle B' + \angle C'$

  H $\angle A + \angle B + \angle C < \angle A' + \angle B' + \angle C'$

  J $\angle A = \angle A',\ \angle B = \angle B',\ \angle C = \angle C'$

Hint

When a figure is reflected, the measures of the angles do not change.

39. Look at the graph below.

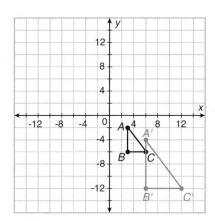

What is the relationship between triangle *ABC* and triangle *A'B'C'*?

A *ABC* and *A'B'C'* are congruent.

B *ABC* and *A'B'C'* are similar.

C *ABC* and *A'B'C'* are mirror images of each other.

D *ABC* and *A'B'C'* are not related to each other in any way.

Hint

In a dilation, the two figures are similar.

40. Look at the graph below.

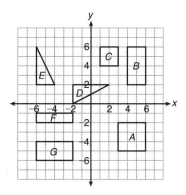

Which one of the following identifies similar figures?

F Figure A is similar to figure C because of dilation and translation.

G Figure A is similar to figure F because of reflection and translation.

H Figure F is similar to figure C because of translation and reflection.

J Figure A is similar to figure G because of translation.

Hint

Similar figures have the same shape but not the same size.

41. Look at the figure below.

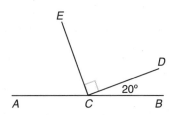

What is the measure of ∠ACE?

A 20° **C** 70°

B 65° **D** 80°

Hint

Look for complementary and supplementary angles.

42. The table below gives the lengths of the three sides of four triangles.

Triangles	Lengths of Sides
F	3, 4, 5
G	9, 12, 15
H	27, 36, 45
J	10, 24, 25

Which triangle is NOT a right triangle?

F △F **H** △H

G △G **J** △J

Hint

Use the formula $a^2 + b^2 = c^2$.

43. If rectangle ABCD has its vertices at points $A = (-7, 7)$, $B = (-7, 2)$, $C = (-4, 2)$, and $D = (-4, 7)$ and is rotated 90° on point C, which one of the following is true about rectangle $A'B'C'D'$?

A The lengths of the respective sides become larger.

B The lengths of the respective sides do not change.

C The area of the rectangle becomes larger.

D The perimeter of $A'B'C'D'$ is twice as large as that of ABCD.

Hint

The figure formed by a rotation is congruent to the original figure.

44. The vertices of rectangle ABCD are $(-4, 2)$, $(-7, 2)$, $(-4, 7)$, and $(-7, 7)$. Which one of the following gives the vertices of rectangle $A'B'C'D'$ if ABCD is reflected across the origin?

F $(2, -4)$, $(2, -7)$, $(7, -4)$, and $(7, -7)$

G $(4, -2)$, $(7, -2)$, $(4, -7)$, and $(7, -7)$

H $(4, -2)$, $(7, -2)$, $(-7, 4)$, and $(-7, 7)$

J $(4, -2)$, $(7, -2)$, $(4, -7)$, and $(-7, 7)$

Hint

Reflections across the origin change the signs of the vertices.

45. Look at the figure below. Line $a \parallel$ line b.

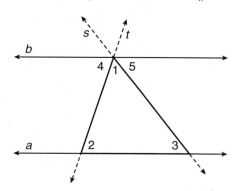

Which one of the following does NOT give the measure of $\angle 1$?

A $\angle 1 = 90 - \angle 3$

B $\angle 1 = 180 - (\angle 4 + \angle 5)$

C $\angle 1 = 180 - (\angle 2 + \angle 3)$

D $\angle 1 = 180 - \angle 3 + \angle 4$

Hint

Use the angle properties of parallel lines and the sum in a triangle.

46. Look at the graph below. Rectangle $E'F'G'H'$ is a dilation of $EFGH$ with the center of dilation at E and a scale factor of $\frac{1}{2}$.

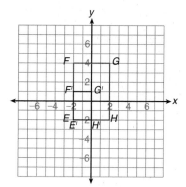

Which one of the following is NOT true?

F $E'F'G'H'$ is congruent to $EFGH$.

G $FG \parallel F'G'$ and $EH \parallel E'H'$.

H The length of $E'H'$ is one half the length of EH.

J $E'F'G'H'$ is similar to $EFGH$.

Hint

A dilation does change the size of an object.

47. Triangle *ABC* is reflected across the origin of a coordinate plane. Which one of the following is NOT true?

A ∠*A* goes to ∠*A'*, ∠*B* goes to ∠*B'*, and ∠*C* goes to ∠*C'*.

B ∠*A* = ∠*A'*, ∠*B* = ∠*B'*, ∠*C* = ∠*C'*

C The sum of the angles in *A'B'C'* is not equal to the sum of the angles of *ABC*.

D The sum of the angles in *ABC* is equal to the sum of the angles in *A'B'C'*.

Hint

A reflection does not change the size of the angles.

48. Look at the figure below. Triangle *ABC* and triangle *DEF* are similar.

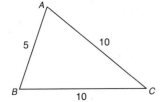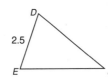

Which proportion would you use to find the length of *EF*?

F $\dfrac{AB}{BC} = \dfrac{BC}{CA}$

H $\dfrac{AB}{AC} = \dfrac{AB}{DE}$

G $\dfrac{DE}{EF} = \dfrac{EF}{DF}$

J Not Here

Hint

Proportions should compare like quantities.

49. Look at the graph below.

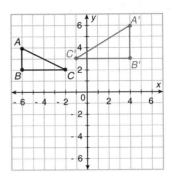

Which one of the following transformations was done to △*ABC* to form △*A'B'C'*?

A translation and rotation

B rotation and dilation

C reflection and dilation

D reflection and rotation

Hint

Rotation changes the orientation, reflection changes the direction, translation moves the figure, and dilation changes the size.

50. Trapezoid *ABCD* is reflected and rotated on a coordinate plane to form *A'B'C'D'*. Which one of the following is true?

F The parallel sides are no longer parallel.

G The lengths of the sides change.

H *ABCD* is congruent to *A'B'C'D'*.

J The measures of the angles change.

Hint

What changes when a figure is reflected and rotated?

Geometry

Independent Practice

DIRECTIONS: Read each question and choose the best answer. Use the answer sheet provided at the end of the workbook to record your answers. If the correct answer is not available, mark the letter for "Not Here."

51. Triangle ABC is translated to triangle $A'B'C'$ by moving it up 1 and over 2 on a coordinate plane. What is always true about the sides of the two triangles?

   A The sides of $A'B'C'$ are larger than the sides of ABC.

   B The sides of $A'B'C'$ are smaller than the sides of ABC.

   C The sides of ABC are equal to the sides of $A'B'C'$.

   D The sides of $A'B'C'$ cross the x-axis.

52. The vertices of right triangle ABC are $A = (-7, 7)$, $B = (-7, 2)$, and $C = (-1, 2)$. What is the length of AC to the nearest whole number?

   F 3   **H** 7

   G 4   **J** 8

53. Rectangle $ABCD$ is transformed into rectangle $A'B'C'D'$ by moving it 3 down and 5 across. What is the transformation called?

   A dilation   **C** rotation

   B reflection   **D** translation

54. Rectangle $ABCD$ has vertices $A = (1, 6)$, $B = (1, 2)$, $C = (7, 2)$, and $D = (7, 6)$. What is the length of a diagonal of the rectangle to the nearest whole number?

   F 3   **H** 7

   G 4   **J** Not Here

55. Look at the graph below.

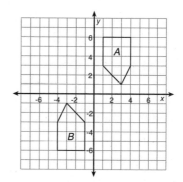

Which combination of transformations changed figure A to figure B?

   A translation followed by rotation

   B reflection followed by reflection

   C reflection over the y-axis followed by translation over the x-axis

   D reflection over x-axis followed by rotation

56. Which statement is true regarding the rotation of an angle?

   F The angle measure increases as it rotates.

   G The direction of the opening of the angle changes as it rotates.

   H The angle measure decreases.

   J Adjacent angles are formed as angles rotate.

57. What is the length of the side of a cone if the height is 25 and the radius of the base is 8? Choose the closest approximation.

A   between 23 and 25

B   between 23 and 24

C   between 26 and 28

D   between 25 and 26

58. The table below gives the lengths of three sides of various triangles.

Triangles	Lengths of Sides
A	7, 9, 12
B	5, 8, 9
C	8, 15, 17
D	40, 50, 80

Which triangle is a right triangle?

F $\triangle A$ H $\triangle C$

G $\triangle B$ J $\triangle D$

59. Look at the figure below. Lines a and b are parallel.

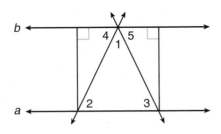

Which one of the following is true?

A $\angle 2 + \angle 4 = 90°$

B $\angle 2 + \angle 3 + \angle 1 = 90°$

C $\angle 3 = \angle 5$

D $\angle 2 = \angle 5$

60. Figure $ABCD$ is dilated with a scale factor of $\frac{2}{3}$ to form figure $A'B'C'D'$. Which one of the following is true?

F   The parallel lines in $ABCD$ are also parallel in $A'B'C'D'$.

G   The angles in $ABCD$ are not equal to the respective angles in $A'B'C'D'$.

H   The figure $ABCD$ is congruent to $A'B'C'D'$.

J   The sides of $ABCD$ are equal to the respective sides of $A'B'C'D'$.

61. If triangle ABC is drawn on a coordinate plane and is reflected over the x-axis, which one of the following is a true statement about triangles ABC and $A'B'C'$?

A   The triangles are similar but not congruent.

B $\angle A = \angle A'$, $\angle B = \angle B'$, and $\angle C = \angle C'$

C $\angle A$ and $\angle A'$, $\angle B$ and $\angle B'$, and $\angle C$ and $\angle C'$ are supplementary.

D   The point $(0, 0)$ will be a vertex of $A'B'C'$.

62. Two triangles are drawn on a coordinate plane. The second triangle is similar to the first, but it is not congruent to it. Which transformation was used to create the second triangle?

F   dilation   H   rotation

G   reflection   J   translation

63. An isosceles triangle has vertices $A = (-5, -1)$, $B = (-7, -6)$, and $C = (-3, -6)$. What is the length of one of the equal sides?

A   3.4   C   4.4

B   3.5   D   5.4

64. Quadrilateral *ABCD* is rotated 90° in a clockwise direction forming *A′B′C′D′*. Which one of the following is true?

F The sum of the angles of *ABCD* is less than the sum of the angles of *A′B′C′D′*.

G The sum of the angles of *ABCD* is greater than the sum of the angles of *A′B′C′D′*.

H The sum of the angles of *A′B′C′D′* equals 180°.

J The sum of the angles of *A′B′C′D′* equals 360°.

65. Rectangle *ABCD* is translated to form rectangle *A′B′C′D′*. What is NOT true about *A′B′C′D′*?

A The lines that were parallel in *ABCD* remain parallel in *A′B′C′D′*.

B The sides of *A′B′C′D′* are longer than the sides of *ABCD*.

C The sides of *A′B′C′D′* are equal to the corresponding sides of *ABCD*.

D *ABCD* is congruent to *A′B′C′D′*.

66. Look at the graph below.

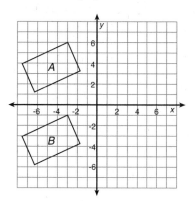

What was done to figure A to form figure B?

F translation

G reflection over *x*-axis

H rotation around the origin

J dilation

67. Look at the graph below.

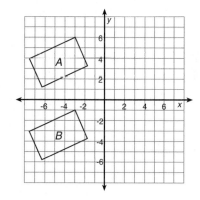

Which one of the following is true of figure A and figure B?

A The figures are similar but not congruent.

B The figures are congruent.

C The area of figure B is greater than the area of figure A.

D The perimeter of figure B is not equal to the perimeter of figure A.

68. What is the volume of a cone if the radius of its base is 6 inches and its height is 15 inches? Use $\pi = 3.14$.

F 94.2 in.^{3}

G 251.2 in.^{3}

H 565.2 in.^{3}

J 1,695.6 in.^{3}

69. Figure *ABCDE* is translated 5 down and 3 to the right to form figure *A'B'C'D'E'*. Which one of the following is a true statement?

A The sum of angles *A*, *C*, and *E* is greater than the sum of angles *A'*, *C'*, and *E'*.

B The sum of the angles in *ABCDE* is less than the sum of the angles in *A'B'C'D'E'*.

C $\angle A = \angle A'$, $\angle B = \angle B'$, $\angle C = \angle C'$, $\angle D = \angle D'$, and $\angle E = \angle E'$

D The sum of the angles of *ABCDE* is greater than the sum of the angles of *A'B'C'D'E'*.

70. Look at the graph below. *ABCD* is reflected and translated to form *A'B'C'D'*.

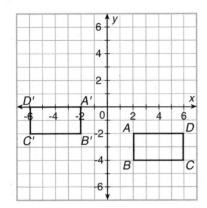

Which one of the following is NOT true?

F $AD = A'D'$ **H** $CD = C'D'$

G $AB \nparallel A'B'$ **J** $AB \parallel A'B'$

71. Look at the graph below.

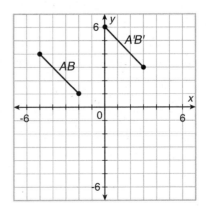

Which one of the following is NOT true?

A *A'B'* was formed by translating *AB* over 5 and up 2.

B $AB = A'B'$

C $AB \nparallel A'B'$

D The slope of *AB* equals the slope of *A'B'*.

72. Look at the figure below.

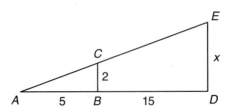

Which proportion can be used to find the length of *DE*?

F $\dfrac{2}{5} = \dfrac{x}{15}$

G $\dfrac{2}{15} = \dfrac{x}{5}$

H $\dfrac{2}{5} = \dfrac{x}{20}$

J $\dfrac{x}{2} = \dfrac{5}{15}$

73. There are two lines on a coordinate plane. *AB* has coordinates ($-$ 1, 2) and (-5, 4). *A'B'* has coordinates (1, 1) and (3, 4). Which one of the following statements is true?

  A   *A'B'* is a reflection of *AB*.

  B   The two lines are parallel.

  C   *A'B'* is a translation of *AB*.

  D   *A'B'* is a rotation of *AB*.

74. Rectangle *ABCD* is rotated 90° counterclockwise to form *A'B'C'D'*. Which one of the following is NOT true?

  F   Corresponding sides of *ABCD* and *A'B'C'D'* are parallel.

  G   *ABCD* is congruent to *A'B'C'D'*.

  H   The ratio of *ABCD* to *A'B'C'D'* is 1.

  J   Opposite sides of *A'B'C'D'* are parallel.

75. The hypotenuse of a right triangle is 12 inches. One of the sides is 9 inches. What is the length of the third side?

  A $\sqrt{7.93}$ in.

  B   9 in.

  C   15 in.

  D $\sqrt{63}$ in.

76. Look at the graph below.

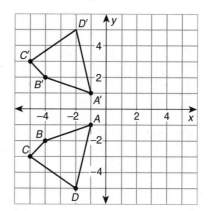

What is NOT true about figures *ABCD* and *A'B'C'D'*?

  F   *A'B'C'D'* is a reflection of *ABCD*.

  G   *ABCD* is congruent to *A'B'C'D'*.

  H   *ABCD* and *A'B'C'D'* are similar but not congruent.

  J   The perimeters of *ABCD* and *A'B'C'D'* are equal.

77. Look at the graph below.

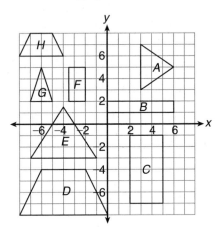

Which one of the following is true?

A Figure D and figure H are similar as a result of dilation and translation.

B Figure B and figure F are similar because of rotation and translation.

C Figure B is congruent to figure C because of reflection and rotation.

D Figure C is similar to figure F as a result of reflection.

78. If a figure is transformed by dilation, what is true about the line segments that form the sides of the figure?

F The line segments remain the same size.

G The size of the line segments changes according to the scale factor.

H Some of the line segments change size, and others do not.

J The line segments change orientation with respect to the x- and y-axes.

79. What is the volume of a cylindrical water tank 20 feet high and 20 feet across? Use $\pi = 3.14$.

A 638 ft³   **C** 8,003.14 ft³

B 1,256 ft³   **D** 6,280 ft³

80. Which one of the following transformations will result in figures that are NOT congruent?

F reflection and rotation

G translation and reflection

H dilation and translation

J two translations

81. A trapezoid $ABCD$ is reflected over the x-axis and then rotated 90° clockwise to form trapezoid $A'B'C'D'$. $AD \parallel BC$ and $AB \nparallel DC$. Which is true?

A $AD \parallel A'D'$ **C** $A'D' \parallel D'C'$

B $A'B' \nparallel D'C'$ **D** $AB \parallel A'B'$

82. Triangle ABC has vertices $A = (-3, -2)$, $B = (-3, -4)$, and $C = (1, -4)$. What is the length of AC?

F $\sqrt{2}$ **H** $\sqrt{12}$

G $\sqrt{6}$ **J** $\sqrt{20}$

83. The coordinates of triangle ABC are $A = (-4, 4)$, $B = (-4, 0)$, and $C = (-1, 2)$. The coordinates of triangle $A'B'C'$ are $A' = (-2, 2)$, $B' = (-2, 0)$, and $C' = (-\frac{1}{2}, 1)$. Which one of the following is true?

A $\triangle A'B'C'$ is a translation of $\triangle ABC$.

B $\triangle A'B'C'$ is congruent to $\triangle ABC$ because it was formed by translation.

C $\triangle ABC$ is similar to $\triangle A'B'C'$ because $\triangle A'B'C'$ is a dilation of $\triangle ABC$.

D $\triangle A'B'C'$ and $\triangle ABC$ are congruent because $\triangle A'B'C'$ was formed by rotation.

84. A ladder 15 feet tall is leaning on the side of a house. The base of the ladder is 6 feet away from the house. How far up on the house is the top of the ladder? Choose the closest approximation.

F between 13 and 14 ft

G between 13.5 and 13.8 ft

H between 13.69 and 13.72 ft

J between 13.70 and 13.80 ft

85. A cylinder and a cone have the same height and the same diameter of the base. If their height is 50 cm and the diameter of the base is 34 cm, which one of the following statements is true? Use $\pi = 3.14$.

A The volumes are equal.

B The volume of the cone is $\frac{1}{3}$ the volume of the cylinder.

C The volume of the cylinder is $\frac{1}{3}$ the volume of the cone.

D Not Here

86. A right triangle has negative x and y coordinates. The triangle is reflected over the y-axis. What happens to the signs of the coordinates?

F The y coordinate becomes positive, and the x coordinate stays negative.

G The signs of the coordinates do not change.

H The x coordinate and y coordinate become positive.

J The y coordinate stays negative, but the x coordinate becomes positive.

87. On a coordinate plane, triangle ABC is congruent to triangle $A'B'C'$. Which one of the following transformations could NOT have happened to form triangle $A'B'C'$?

A rotation, reflection, translation

B rotation, reflection, dilation

C reflection, translation, rotation

D translation, reflection, rotation

88. A rectangular box is 8 ft by 8 ft by 4 ft. What is the longest drapery rod that can be put in the box?

F 8.9 ft   **H** 12 ft

G 11.31 ft   **J** 14.32 ft

89. The table below gives the lengths of the three sides of four triangles. Which triangle is NOT a right triangle?

Triangle	Lengths of Sides
A	3, 4, 5
B	9, 12, 15
C	40, 50, 80
D	8, 15, 17

A △A

B △B

C △C

D △D

90. Parallelogram $A'B'C'D'$ is formed by dilation of parallelogram $ABCD$ with a scale factor of 2. Which one of the following is NOT true?

F The sum of the angles of $ABCD$ is equal to the sum of the angles of $A'B'C'D'$.

G The angles of $A'B'C'D'$ are equal to the respective angles of $ABCD$.

H The opposite angles of $A'B'C'D'$ are equal to the respective opposite angles of $ABCD$.

J The sides of $A'B'C'D'$ are equal to the respective sides of $ABCD$.

91. Look at the graph below. Triangle $A'B'C'$ is a dilation of triangle ABC.

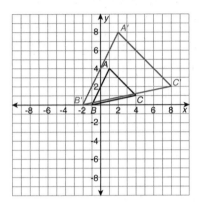

Which one of the following is true?

A △ABC is similar to △$A'B'C'$.

B △ABC is congruent to △$A'B'C'$.

C The scale factor is $\frac{1}{2}$.

D The angles of △ABC are not equal to the respective angles of △$A'B'C'$.

92. The sides of a triangle are 15 cm, 17 cm, and 19 cm. Is the triangle a right triangle? Why or why not?

F No, $15^2 + 17^2 \neq 19^2$.

G Yes, $15^2 + 17^2 = 19^2$.

H No, $19^2 + 17^2 \neq 15^2$.

J Yes, $19^2 + 15^2 = 17^2$.

93. Look at the graph below. Figure *ABC* is translated to form figure *A'B'C'* by moving *ABC* 4 down and 3 to the right.

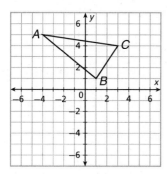

Which one of the following is true?

A $AC \parallel BC$, $BC \parallel B'C'$, $AB \parallel A'B'$

B $AC \nparallel A'C'$, $BC \nparallel B'C'$, $AB \parallel A'B'$

C $AC \parallel A'C'$, $BC \parallel B'C'$, $AB \parallel A'B'$

D $\angle A = \angle A'$, $\angle B = \angle C'$, $\angle C = \angle B'$

94. Two parallel lines are drawn on a coordinate plane, each with a slope of -4. Which one of the following is true if the lines are reflected over the *x*-axis?

F The slope of the reflected lines will be 4.

G The reflected lines will be closer to each other than the original lines.

H The slope of the reflected lines will be $-\dfrac{1}{4}$.

J The slope of the reflected lines will be -4.

95. In the figure below, $AB \parallel CD$.

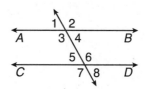

Which one of the following lists pairs of supplementary angles?

A $\angle 2$ and $\angle 8$, $\angle 1$ and $\angle 7$, $\angle 3$ and $\angle 6$, $\angle 4$ and $\angle 5$

B $\angle 2$ and $\angle 3$, $\angle 6$ and $\angle 7$, $\angle 5$ and $\angle 8$, $\angle 1$ and $\angle 4$

C $\angle 3$ and $\angle 4$, $\angle 5$ and $\angle 6$, $\angle 6$ and $\angle 7$, $\angle 1$ and $\angle 2$

D $\angle 2$ and $\angle 8$, $\angle 1$ and $\angle 7$, $\angle 3$ and $\angle 5$, $\angle 6$ and $\angle 4$

96. Which one of the following transformations will NOT result in congruent figures?

F dilation   **H** rotation

G reflection   **J** translation

97. The vertices of triangle *ABC* are $A = (7, -6)$, $B = (4, -2)$, and $C = (-7, -2)$. Triangle *A'B'C'* is reflected across the *y*-axis. Which one of the following statements is true?

A $\angle A + \angle B + \angle C < \angle A' + \angle B' + \angle C'$

B $\angle A = \angle A'$, $\angle B = \angle B'$, $\angle C = \angle C'$

C $\angle A = \angle A'$, $\angle B < \angle B'$, $\angle C > \angle C'$

D $\angle A + \angle B + \angle C > \angle A' + \angle B' + \angle C'$

Name _____ **Date** _____

98. How many boxes that are 5 in. × 5 in. × 10 in. will fit in a shipping carton that is 12 in. × 12 in. × 20 in.?

F 7   **H** 8

G 7.5   **J** 8.5

99. What are the coordinates of the image of the point $(11, 1)$ after a translation 10 units down followed by a reflection over the x-axis?

A $(11, -9)$ **C** $(-11, 9)$

B $(11, 9)$ **D** $(-11, -9)$

100. The coordinates of triangle ABC are $A = (-4, 4)$, $B = (-6, 0)$, and $C = (-1, 0)$. What are the coordinates of dilation with a scale factor of $\frac{1}{2}$?

F $A = (-3.5, 3.5)$, $B = (-5.5, -0.5)$, $C = (-1.5, -0.5)$

G $A = (-4, 2)$, $B = (-3, 0)$, $C = (-\frac{1}{2}, 0)$

H $A = (-2, 2)$, $B = (-3, 0)$, $C = (-\frac{1}{2}, 0)$

J $A = (-2, 4)$, $B = (-3, 0)$, $C = (-\frac{1}{2}, 0)$

101. Triangle ABC is rotated 90° clockwise to form triangle $A'B'C'$. Which one of the following is NOT true?

A The sum of the angles of ABC equals the sum of the angles of $A'B'C'$.

B The sum of the angles of ABC is not equal to the sum of the angles of $A'B'C'$.

C $\angle A$ goes to $\angle A'$, $\angle B$ goes to $\angle B'$, and $\angle C$ goes to $\angle C'$.

D $\angle A = \angle A'$, $\angle B = \angle B'$, and $\angle C = \angle C'$.

102. Triangle $A'B'C'$ is a dilation of triangle ABC with a scale factor of $\frac{1}{2}$. Which one of the following is true of the two triangles?

F $AB = A'B'$, $BC = B'C'$, $BA = B'A'$

G $\angle A' + \angle B' + \angle C' = 360°$

H $\angle A = \angle A'$, $\angle B = \angle B'$, $\angle C = \angle C'$

J The triangles are congruent.

103. The coordinates of triangle ABC are $A = (-2, 4)$, $B = (-2, 1)$, and $C = (0, 1)$. What are the coordinates of $A'B'C'$ if it is reflected over the y-axis?

A $A' = (2, 4)$, $B' = (2, 1)$, and $C' = (0, 1)$

B $A' = (4, 2)$, $B' = (1, 2)$, and $C' = (1, 0)$

C $A' = (-2, -4)$, $B' = (-2, -1)$, and $C' = (0, -1)$

D $A' = (2, -4)$, $B' = (2, -1)$, and $C' = (0, -1)$

104. A parallelogram has vertices $A = (-1, 2)$, $B = (-3, -3)$, $C = (5, -3)$, and $D = (7, 2)$. A line perpendicular to BC is drawn from A to BC. What is the length of AB?

F $\sqrt{7}$

G $\sqrt{21}$

H $\sqrt{23}$

J $\sqrt{29}$

105. Look at the graph below.

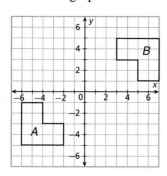

What series of transformations took place to create figure B from figure A?

A   dilation and translation

B   reflection and rotation

C   translation and translation

D   rotation and translation

106. In the figure below, line a is parallel to line b.

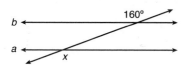

What is the measure of $\angle x$?

F   20°   **H**   140°

G   40°   **J**   160°

107. What is the volume of a ball with a diameter of 18 inches? Use $\pi = 3.14$.

A   37.68 in.³   **C**   3,052.08 in.³

B   75.36 in.³   **D**   7,776 in.³

108. Two figures are drawn on a coordinate plane. The two figures are not congruent. Which transformation could have been used to create the second figure?

F   reflection, translation, rotation

G   dilation, reflection, translation

H   translation, reflection, reflection

J   reflection, rotation, translation

109. A quadrilateral has vertices $A = (-4, 5)$, $B = (-4, 1)$, $C = (4, 1)$, and $D = (0, 5)$. What is the length of side DC to the nearest tenth?

A   2.8   **C**   3.4

B   3.1   **D**   5.7

110. Look at the figure below.

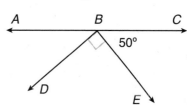

What is the measure of $\angle ABD$?

F   40°   **H**   90°

G   50°   **J**   180° − 50°

Name _____ Date _____

Statistics and Probability

Modeled Instruction

DIRECTIONS: Read each question and choose the best answer. Use the answer sheet provided at the end of the workbook to record your answers. If the correct answer is not available, mark the letter for "Not Here."

1. The trend line on a graph goes from upper left to lower right. What does this show about the data?

   A Both data sets increase together.

   B As one set of data increases, the other decreases.

   C A change in one set of data has no effect on the other set.

   D The data are not accurate.

   Hint

   Sketch a graph to help you.

2. Look at the graph below.

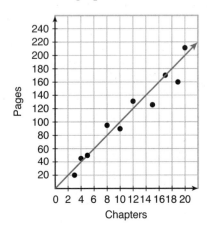

   What is the equation of the trend line in the graph?

   F $3y - 27x + 2$ **H** $y = 10x$

   G $y = 9x - 2$ **J** $27x = 3y - 2$

   Hint

   $y = mx + b$, and slope $= \dfrac{y_2 - y_1}{x_2 - x_1}$.

Name _____ Date _____

3. Look at the data table below.

Chapters	1	7	9	13	17
Pages	2	14	18	26	34

Which one of the following is the best description for the data?

A positive association, slope: $\frac{2}{1}$, equation: $y = 2x + 10.5$

B positive association, slope: $\frac{1}{2}$, equation: $y = \frac{1}{2}x - 10.5$

C negative association, slope: $\frac{2}{1}$, equation: $y = 2x + 10.5$

D negative association, slope: $\frac{1}{2}$, equation: $y = \frac{1}{2}x + 10.5$

Hint

Look at the direction for association. Slope equals $\frac{y_2 - y_1}{x_2 - x_1}$. The equation is $y = mx + b$.

---

4. The recreation department is offering swimming lessons for students ages 6, 7, and 8 who can't swim. The table below shows results of a survey that was taken to determine how many students would be taking lessons.

	Swimmers	Non-Swimmers	Total
6	50	10	60
7	73	8	81
8	80	5	85
Total	203	23	226

What is the relative frequency of 8-year-olds who can swim?

F 2% **G** 35% **H** 37% **J** 89%

Hint

Relative frequency is the ratio of the number to the total.

5. Look at the data table below.

Hours in Mall	9	1	5	7	2	8
Dollars Spent ($)	24	10	50	18	100	60

Which scatter plot below shows the points for these data?

A

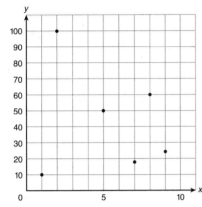

C

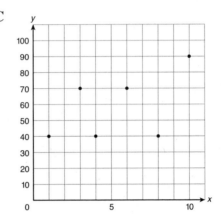

B

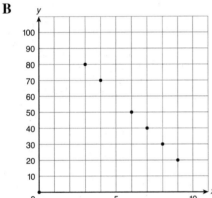

D

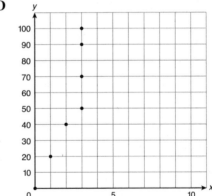

Hint

Compare the data chart with the points in the scatter plots.

Statistics and Probability
Higher Scores on Math, Grade 8

Name _____   Date _____

6. Look at the graph below.

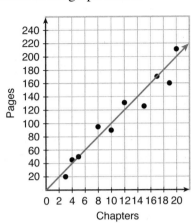

Which one of the following is the best guess for the number of pages in 3.5 chapters?

F 15   **G** 20   **H** 35   **J** Not Here

Hint

Find the value of y when $x = 3.5$.

---

7. Which one of the following gives the slope and equation of a trend line that goes through the points $(3, 7)$ and $(10, 2)$? What kind of association does the line show?

A slope: $-\dfrac{7}{5}$, equation: $y = -\dfrac{7}{5}x + 64$, negative association

B slope: $-\dfrac{5}{7}$, equation: $y = -\dfrac{5}{7}x + \dfrac{64}{7}$, negative association

C slope: $\dfrac{5}{7}$, equation: $y = \dfrac{5}{7}x + \dfrac{64}{7}$, positive association

D slope: $\dfrac{7}{5}$, equation: $y = \dfrac{7}{5}x + 64$, positive association

Hint

$y = mx + b$ and slope $= \dfrac{y_2 - y_1}{x_2 - x_1}$.

Name _____   Date _____

8. A political party is conducting a survey of which candidate is preferred. Results are categorized by age group of the responders. The results appear in the table below.

	Candidate A	Candidate B	Candidate C	Total
18 to 30 years	500	763	922	2,185
31 to 60 years	840	350	200	1,390
61 and over	1,000	375	800	2,175
Total	2,340	1,488	1,922	5,750

What is the relative frequency of the 61-and-over group who responded to the survey?

F 17%   **G** 24%   **H** 38%   **J** 62%

Hint

Relative frequency is the ratio of the number to the total.

Statistics and Probability

Independent Practice

DIRECTIONS: Read each question and choose the best answer. Use the answer sheet provided at the end of the workbook to record your answers. If the correct answer is not available, mark the letter for "Not Here."

9. A survey group is conducting a survey to determine the most popular TV channel during prime time. The survey results are by age group. The results appear below.

	Channel 3	Channel 7	Channel 9	Total
18 to 30 years	273	540	427	1,240
31 to 60 years	115	217	320	652
61 and older	407	619	725	1,751
Total	795	1,376	1,472	3,643

What is the relative frequency of the 61-and-older group that preferred Channel 9?

A 8.7 %   **B** 16%   **C** 19.9%   **D** 210%

---

10. Look at the graph below.

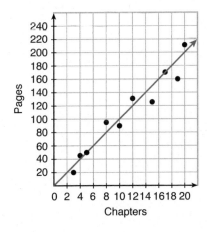

Which one of the following gives the best guess of pages for 16 chapters?

F 140   **G** 150   **H** 155   **J** 160

11. Look at the scatter plot below.

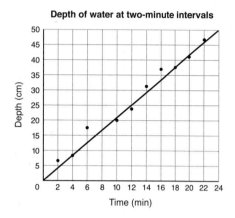

Depth of water at two-minute intervals

What are the slope and equation of the trend line?

A slope: 2, equation: $y = 2x$

B slope: -2, equation: $y = -2x$

C slope: $\frac{1}{2}$, equation: $y = \frac{1}{2}x$

D slope: $-\frac{1}{2}$, equation: $y = -\frac{1}{2}x$

12. Look at the scatter plot below.

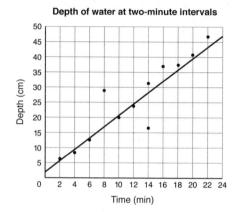

Depth of water at two-minute intervals

Which one of the following is the best description of the scatter plot?

F negative association, slope: -2, equation: $y = -2x + 1$

G negative association, slope: $-\frac{1}{2}$, equation: $y = -\frac{1}{2}x + 1$

H positive association, slope: 2, equation: $y = 2x + 1$

J positive association, slope: $\frac{1}{2}$, equation: $y = \frac{1}{2}x + 1$

13. Look at the graph below.

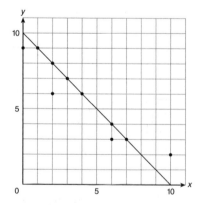

What is the equation of the trend line?

A $y = x - 5$

C $y = -x + 10$

B $3y = 3x - 15$

D $y = -x + 5$

14. Look at the graph below.

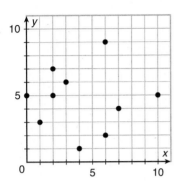

Which one of the following describes the graph?

F outlier

G negative association

H no association

J cluster

---

15. Look at the table below.

	Soccer	**Basketball**
Football	Football and Soccer	a
Baseball	b	c

Which one of the following gives the correct replacements for a, b, and c in the table?

A

a	b	c
Baseball and Basketball	Baseball and Soccer	Football and Basketball

B

a	b	c
Baseball and Football	Baseball and Basketball	Football and Soccer

C

a	b	c
Football and Soccer	Baseball and Basketball	Football and Baseball

D

a	b	c
Football and Basketball	Baseball and Soccer	Baseball and Basketball

16. Look at the table below.

	Soccer	Baseball	Total
Girls	20	15	35
Boys	18	18	36
Total	38	33	71

Compare the relative frequencies of girls who like baseball to boys who like baseball.

F   Boys who like baseball is greater by 4%.

G   Girls who like baseball is greater by 4%.

H   Boys who like baseball is the same as girls who like baseball.

J   Boys who like baseball is less than girls who like baseball by 3%.

17. The scatter plot shows the number of students and the number of hours of homework they do in a week.

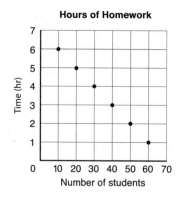

Hours of Homework

How many more students do 2 hours of homework than 4 hours of homework?

A   10   **C**   30

B   20   **D**   40

18. Look at the table below.

	Chores	No Chores	Total
Curfew	60	28	88
No Curfew	60	10	70
Total	120	38	158

Which conclusion CANNOT be drawn from this table?

F   The relative frequency of students doing chores to total surveyed is 76%.

G   There is no relationship between a curfew and doing chores.

H   Students with no curfew also do the fewest chores.

J   The relative frequency of students with no curfew to the total is 23%.

19. Look at the graph below comparing the depth of the water to the fill time.

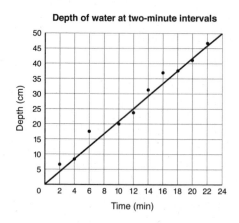

Depth of water at two-minute intervals

Which one of the following is the best guess for 8 minutes?

A   15   **C**   19

B   17   **D**   20

Name _____   Date _____

20. Look at the table below.

Price ($)	5	10	15	20	25
Buyers	30	20	10	5	3

Which scatter plot is the best representation of the data?

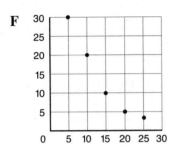

F

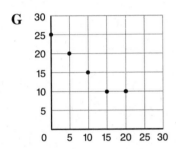

G

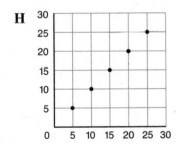

H

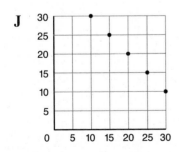

J

21. Look at the scatter plot below.

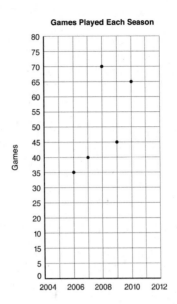

Games Played Each Season

How many more games were played in 2006 and 2007 combined than in 2008?

A   4

B   5

C   6

D   The number of games is the same.

22. What is a set of data that involves two variables called?

F   scatter plot data

G   bivariate data

H   negative association data

J   positive association data

23. In which one of the following is a trend line most helpful?

    A   no association

    B   cluster

    C   linear association

    D   nonlinear association

24. The scatter plot and trend line on a graph go from upper right to lower left. What does this indicate about the data?

    F   The data are not accurate.

    G   A change in one set of data does not affect the other set.

    H   As one set of data increases, the other decreases.

    J   Both sets of data increase together.

Name _____   Date _____

Practice Test A

DIRECTIONS: Read each question and choose the best answer. Use the answer sheet provided at the end of the workbook to record your answers. If the correct answer is not available, mark the letter for "Not Here."

1. In the figure below, $AB \parallel CD$.

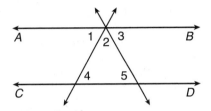

Which one of the following is true?

A $\angle 4 = \angle 2$ and $\angle 5 = \angle 3$

B $\angle 4 = \angle 1$ and $\angle 5 = \angle 3$

C $\angle 4 + \angle 2 + \angle 5 = 90°$

D $\angle 5 + \angle 1 = 90°$

2. A cube has a volume of 19,683 cm³. What is the length of each side?

F 27 cm

G 140 cm

H 729 cm

J 6,561 cm

3. Triangle $A'B'C'$ was formed by a dilation of triangle ABC with a scale factor of $\frac{1}{2}$. What is true about the relationship between segments AB and $A'B'$, BC and $B'C'$, and AC and $A'C'$?

A AB equals $A'B'$, BC equals $B'C'$, and AC equals $A'C'$.

B AB is half the length of $A'B'$, BC is half the length of $B'C'$, and AC equals $A'C'$.

C Each of the lines in $A'B'C'$ is half the length of the corresponding lines of ABC.

D The two shorter lines of $A'B'C'$ are longer than the two shorter lines of ABC.

4. Evaluate the expression.

$$1.271 \times 10^3 \div 2 \times 10^8$$

What is the correct answer in standard notation?

F 0.000006355

G 0.00006355

H 63,550

J 635,500

5. Look at the scatter plot below.

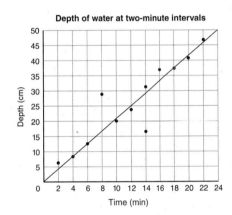

What are the slope and equation of the trend line?

A slope: 2, equation: $y = 2x$

B slope: $\frac{1}{2}$, equation: $y = \frac{1}{2}x$

C slope: -2, equation: $y = -2x$

D slope: $-\frac{1}{2}$, equation: $y = -\frac{1}{2}x$

6. Figure *ABCDE* is drawn on a coordinate graph. Figure *A'B'C'D'E'* is formed through a series of transformations of *ABCDE*. Which one of the following must be true of the two figures if one of the transformations is a dilation?

F The two figures are congruent.

G The lines that were parallel before the transformation are parallel after the transformation.

H The figures are similar but are not congruent.

J The sum of the corresponding angles in each figure is 180°.

7. The scatter plot shows the hours of homework students do in a week.

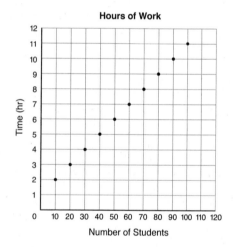

Hours of Work

How many more students do 10 hours of homework than do 6 hours of homework?

A 40 students

B 50 students

C 70 students

D 90 students

8. The equations below define two different functions.

Function 1: $y = 3x + 5$

Function 2: $y = -2x - 3$

What are the rates of change for the two functions, and which has the greater change?

F

Function 1	Function 2	Function with Greater Rate of Change
3	2	Function 1

G

Function 1	Function 2	Function with Greater Rate of Change
3	−2	Function 2

H

Function 1	Function 2	Function with Greater Rate of Change
3	−2	Function 1

J

Function 1	Function 2	Function with Greater Rate of Change
3	2	Function 2

9. Identify the slope and *y*-intercept of the equation $5x + 8y = 40$.

A

Slope	Intercept
$-\dfrac{5}{8}$	-5

C

Slope	Intercept
$-\dfrac{5}{8}$	5

B

Slope	Intercept
$\dfrac{5}{8}$	-5

D

Slope	Intercept
$\dfrac{5}{8}$	5

---

10. In which one of the following is a trend line NOT helpful?

F The data are a linear association.

G The data cluster around a line.

H The data indicate a nonlinear association.

J The data indicate a linear association.

---

11. Look at the table below.

$\sqrt{17}$	$0.666\ldots$	π	0	$-\sqrt{4}$
a	*b*	*c*	*d*	*e*

Classify the numbers as rational or irrational by substituting the correct terms for *a*, *b*, *c*, *d*, and *e*.

A

a	*b*	*c*	*d*	*e*
irrational	irrational	rational	irrational	irrational

B

a	*b*	*c*	*d*	*e*
rational	irrational	rational	rational	irrational

C

a	*b*	*c*	*d*	*e*
irrational	rational	irrational	rational	rational

D

a	*b*	*c*	*d*	*e*
irrational	irrational	irrational	rational	irrational

Practice Test A
Higher Scores on Math, Grade 8

12. If line segment AB has endpoints $(5, 1)$ and $(1, 3)$ and line segment $A'B'$ has endpoints $(-1, 3)$ and $(-5, 1)$, how was $A'B'$ formed?

   F reflection over x-axis

   G reflection over y-axis

   H translation

   J rotation

13. What one characteristic of the original figure remains the same in the figure formed by a translation, reflection, rotation, or dilation either by itself or in combination?

   A The sides remain the same size, but the angles change size.

   B All angles and all sides remain the same size.

   C All angles remain the same size.

   D All sides remain the same size.

14. Parallelogram $A'B'C'D'$ is formed by a dilation of parallelogram $ABCD$ with a scale factor of $\frac{1}{3}$. Which one of the following is a true statement?

   F $A'B' \nparallel D'C', AB \parallel C'D', B'C' \parallel AD$

   G $A'D' \parallel B'C', A'D' \nparallel BC, A'B' \nparallel BC$

   H $AB \parallel A'B', A'B' \parallel D'C', A'B' \parallel CD$

   J $AB \parallel BC, A'B' \parallel B'C', CD \parallel C'D'$

15. Look at the graph below that shows student attendance at school activities during the year.

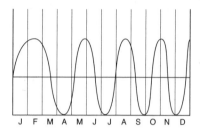

Which of the situations is the best fit for the graph?

   A Attendance at school activities was higher in 31-day months than in 30-day months.

   B Attendance at school activities was lower in 31-day months than in 30-day months.

   C During the shortest month, attendance was high, but dropping.

   D It is impossible to predict an attendance trend.

16. Evaluate the expression.

$$(-3^2)^2 + (4 - 15)^3 \times 5^{-2}$$

   F. 2.1

   G 27.76

   H 29.48

   J 210

17. The local tech store is conducting a survey of customers to determine which group prefers operating system 1 versus operating system 2 on their smart phones. The results are recorded in the table below.

	OS1	OS2	Total
18 – 30	200	40	240
31 – 60	100	125	225
61 and older	50	300	350
Total	350	465	815

What is the difference in relative frequencies between the largest group for OS1 and the largest group for OS2?

A 12.27%

C 36.8%

B 24.5%

D 61%

18. Which one of the following does NOT have the same solution as $35(3 - x) - 20(2 - 3x) = 150$?

F $105 - 35x - 40 + 60x = 150$

G $65 + 25x = 150$

H $105 - x - 40 + 3x = 150$

J $25x = 150 - 65$

19. What figure has the formula $\frac{1}{3}\pi r^2 h$ for finding its volume?

A cone

C prism

B cylinder

D pyramid

20. Parallelogram $ABCD$ on a coordinate plane has vertices $A = (-2, 2)$, $B = (-4, -2)$, $C = (2, -2)$, and $D = (4, 2)$. What are the coordinates of $A'B'C'D'$ if it is a dilation of $ABCD$ with a scale factor of $\frac{1}{2}$?

F $A' = (-2.5, 2.5)$, $B' = (-4.5, -2.5)$,
$C' = (2.5, -2.5)$, and $D' = (4.5, 2.5)$

G $A' = (-4, 2)$, $B' = (-8, -4)$,
$C' = (8, -8)$, and $D' = (8, 4)$

H $A' = (-1, 1)$, $B' = (-2, -1)$,
$C' = (1, -1)$, and $D' = (2, 1)$

J $A' = (1, -1)$, $B' = (2, 1)$, $C' = (-1, 1)$,
and $D' = (-1, -2)$

21. Which of the data tables below gives some of the solutions for the equation $y = x + 3(3x - 9)$?

A

x	2	4	6	8
y	−7	13	33	43

B

x	1	3	5	7
y	−17	3	23	43

C

x	1	4	7	10
y	−17	13	43	75

D

x	2	7	12	17
y	−7	43	93	105

22. Which one of the following is the closest approximation of $\sqrt{51}$?

F between 7 and 8

G between 7 and 7.7

H between 7.1 and 7.2

J between 7.1 and 7.5

23. Look at this system of equations.

$7x + 3y = -6$
$7x + 3y = 5$

Which one of the following best describes the solution?

A all real numbers

B $x = 0, y = 0$

C There is no solution.

D There are an infinite number of solutions.

24. Which one of the following ordered pairs does NOT define a function?

F (5, 7), (6, 9), (7, 10), (7, 12)

G (2, 6), (5, 6), (7, 21), (9, 43)

H (10, 5), (8, 4), (7, 3.5), (5, 2.5)

J (2, 8), (3, 27), (9, 27), (11, 33)

25. Which of the tables below describes a nonlinear function?

A

4	6	7	8
13	21	25	29

B

1	2	3	4
1	4	9	16

C

0	2	3	4
5	9	11	13

D

−3	−2	0	2
−17	−12	−2	8

26. The graph below shows two similar triangles, *ABC* and *A'B'C'*.

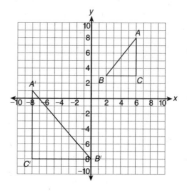

What is the relationship between *AB* and *A'B'*?

F The slopes are equal, but one is positive and the other is negative.

G Both slopes are negative.

H The slopes are the same.

J The slopes are both positive but are not equal.

Name _____ Date _____

27. Look at the scatter plot below.

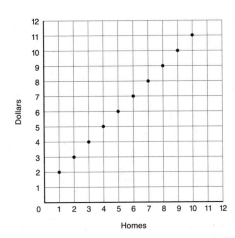

Homes

Which one of the following data charts was used to make the scatter plot?

A

1	2	3	4	5	6	7	8	9	10
2	3	6	8	9	10	11	12	12	12

B

1	2	3	4	5	6	7	8	9	10
2	4	5	6	7	8	9	10	11	11

C

1	2	3	4	5	6	7	8	9	10
2	3	4	5	6	7	8	9	10	11

D

1	4	5	6	8	9	10	11	12	13
1	2	3	4	5	6	7	8	9	11

28. Look at the figure below.

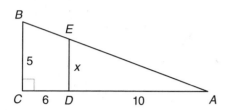

Which proportion can be used to find the length of *DE*?

F $\dfrac{ED}{AD} = \dfrac{EA}{DC}$

H $\dfrac{AE}{ED} = \dfrac{EB}{AC}$

G $\dfrac{ED}{AD} = \dfrac{5}{16}$

J $\dfrac{BC}{DC} = \dfrac{AE}{ED}$

29. What is the length of the side of a cone if the height of the cone is 25 cm and the diameter of the base is 24 cm? Choose the closest approximation.

A between 27.6 and 27.7

B between 17 and 28

C between 27.7 and 27.8

D between 27.9 and 28

30. Solve the system of equations.

$y = 3x + 1$
$4y = x + 3$

Which one of the following is true of the solution?

F The solution is one ordered pair.

G The solution is two ordered pairs.

H There is no solution.

J There are an infinite number of solutions.

31. Trapezoid *ABCD* , with $AD \parallel BC$, is reflected over the *y*-axis and is then translated 3 up and 2 across to form $A'B'C'D'$. Which one of the following is true?

A $AB \parallel DC$ and $A'B' \parallel D'C'$

B $AD \parallel B'C'$ and $BC \parallel A'D'$

C $AD \parallel B'C'$ and $A'B' \parallel C'D'$

D $AD \parallel D'C'$ and $A'B' \parallel AB$

32. The table below gives the lengths of the sides of four triangles.

Triangles	Lengths of Sides
A	3, 4, 5
B	7, 8, 9
C	6, 8, 24
D	13, 17, 19

Which triangle is a right triangle?

F △*A*

G △*B*

H △*C*

J △*D*

33. Two coordinates of line *A* are (6, 7) and (1, 7). Two coordinates of line *B* are (4, 1), and (−5, 1). What is true about lines *A* and *B*?

A They intersect in one point.

B They intersect in two points.

C They are parallel.

D They are the same line.

34. Lilly is shopping for boxes of pens and pencils. She buys a total of 10 boxes and spends $25.00. The pens cost $2.50 per box, and the pencils cost $1.50 per box. How many of each box does she buy?

F 10 boxes of pens and 0 boxes of pencils

G 9 boxes of pens and 1 box of pencils

H 8 boxes of pens and 2 boxes of pencils

J 7 boxes of pens and 3 boxes of pencils

35. Solve the equation.

$$3x + 56 + 3(x + 1) = 3(2x - 9) + 4^2$$

How many solutions does the equation have?

A 0

B 1

C 2

D infinite number

36. Find the volume of a sphere with a radius of 25 cm. Use $\pi = 3.14$.

F 13.08 cm³

G 65416.66 cm³

H 2,616 $\frac{2}{3}$ cm³

J 10,466.66 cm³

37. Look at the graph below showing two linear functions.

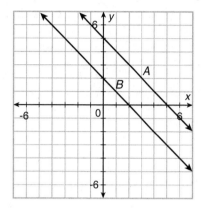

Which one of the following is true?

A Line A has a greater rate of change than Line B.

B Line B has a greater rate of change than Line A.

C The rates of change are the same and equal to -1.

D The rates of change are the same and equal to 1.

38. Javier gets paid $200.00 per week, plus a commission of $15.00 for every new customer he brings to the business. Which equation does NOT show the relationship between the number of new customers and his weekly income?

F $y = 15x + 200$

G $\frac{1}{5}y = 3x + 40$

H $2y = 30x + 200$

J $15x = y - 200$

39. Look at the graph below.

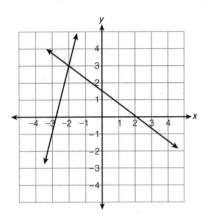

Which ordered pair below is the solution to the system of equations represented by the two lines?

A (2, 0)

B (−2, 3)

C (0, 1.5)

D (−2.75, 0)

40. Trapezoid *ABCD* is rotated 90° counterclockwise forming trapezoid $A'B'C'D'$. Which one of the following is true?

F The sum of the angles of *ABCD* is greater than the sum of the angles of $A'B'C'D'$.

G The sum of the angles of *ABCD* is less than the sum of the angles of $A'B'C'D'$.

H The sum of the angles of $A'B'C'D'$ equals 180°.

J The sum of the angles of $A'B'C'D'$ equals 360°.

41. Which graph below shows a system of equations with no solutions?

A

B

C

D

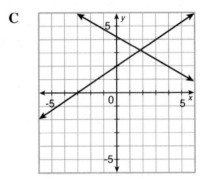

42. The vertices of a right triangle are $A = (-1, -1)$, $B = (-1, -5)$, and $C = (4, -5)$. What is the length of AC to the nearest whole number?

F 4   **H** 6

G 5   **J** 7

43. Rectangle *ABCD* with vertices $A = (-5, 5)$, $B = (-5, 1)$, $C = (-3, 1)$, and $D = (-3, 5)$ is rotated and translated to form $A'B'C'D'$ with vertices $A' = (6, 4)$, $B' = (2, 4)$, $C' = (2, 2)$, and $D' = (6, 2)$. Which one of the following is true?

A $AD = A'D'$

B *ABCD* is similar but not congruent to $A'B'C'D'$.

C $CD > C'D'$

D $AB = A'D'$

44. The graph below represents a function.

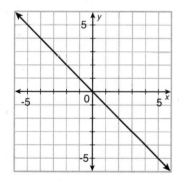

Which equation describes the function?

F $y = x + 1$ **H** $y = x$

G $y = -x + 1$ **J** $y = -x$

45. Look at the scatter plot below.

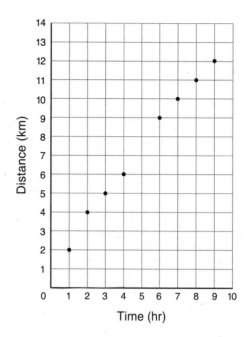

Which is the best guess for how much distance is covered in 5 hours?

A 7 km   **C** 7.5 km

B 7.25 km   **D** 8 km

46. Four times a number is equal to 8 less than 6 times the number. What is the number?

F -4 **H** 2

G -2 **J** 4

47. Look at the graph below.

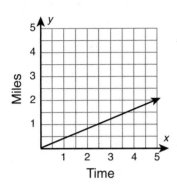

What are the rate of change and slope for the line on the graph?

A rate of change: 2.5, slope: 0.4

B rate of change: 0.4, slope: 0.4

C rate of change: 0.4, slope: 2.5

D rate of change: 2.5, slope: 2.5

48. Sales in the school bookstore are high at the beginning of each semester. Based on the graph below, when does each semester begin?

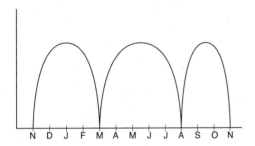

F February, March, and April

G July, March, and April

H December, April, and September

J January, December, and March

49. Which of the equations below has an infinite number of solutions?

A $3x - 2^2 = 4x + 2^3 - 3$

B $2x^2 + 4x - 10 = -16 + 4(1 + x) + 2x^2$

C $x^2 + 2x - 5 = -7 + 2(1 + x) + x^2$

D $3x - 9 + 2(x^2 + 5) = 2x^2 + 3(x - 1)$

50. Compare the following functions, one defined by an equation and one defined by a data table.

Function 1: $y = 3x + 5$
Function 2:

x	2	5	7	9
y	5	17	25	33

What are the rates of change, and which has the greater rate?

F

Function 1	Function 2	Function with Greater Rate of Change
3	4	Function 2

G

Function 1	Function 2	Function with Greater Rate of Change
3	4	Function 1

H

Function 1	Function 2	Function with Greater Rate of Change
−3	4	Function 2

J

Function 1	Function 2	Function with Greater Rate of Change
−3	4	Function 1

51. Evaluate the expression and put the answer in scientific notation.

$(5.67652 \times 10^{15}) \times (4 \times 10^7)$

A 0.2270608×10^{24}

B 22.70608×10^{22}

C 2.270608×10^{8}

D 2.270608×10^{23}

52. The endpoints of line segment AB are $(2, -5)$ and $(5, -2)$. What are the coordinates of segment $A'B'$ if AB is reflected over the origin?

F $(2, 5)$ and $(-5, 2)$

G $(2, -5)$ and $(5, -2)$

H $(-1, -2)$ and $(8, -1)$

J $(-5, 2)$ and $(-2, 5)$

53. What is $(3.69 \times 10^7)^2$ in scientific notation?

A 3.69×10^9 **C** 13.6161×10^{14}

B 1.36161×10^{15} **D** 3.69×10^{14}

54. The coordinates of triangle ABC are $A = (-4, 4)$, $B = (-4, -2)$, and $C = (2, -2)$. The coordinates of triangle $A'B'C'$ are $A' = (-2, 2)$, $B' = (-2, -1)$, and $C' = (1, -1)$. Which one of the following is true?

F $\triangle ABC$ is similar to $\triangle A'B'C'$ because a dilation was performed on $\triangle ABC$.

G The two triangles are congruent because a reflection was performed on $\triangle ABC$.

H $\triangle ABC$ is similar because a rotation was performed on $\triangle ABC$.

J The two triangles are congruent because a translation was performed on $\triangle ABC$.

55. Look at the table below.

3	7	16	40	50
16	28	a	127	b

What are the values of a and b that will make this a function?

A $a = 49, b = 150$ **C** $a = 48, b = 157$

B $a = 55, b = 157$ **D** $a = 50, b = 150$

56. Look at the graph below.

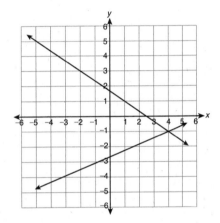

Which ordered pair is the solution to the system of equations represented by the lines on the graph?

F $(3, 0)$ **H** $(0, -3)$

G $(4, -1)$ **J** $(0, 2)$

57. Triangle ABC has vertices $A = (-6, 4)$, $B = (-6, 2)$, $C = (-2, 2)$. Triangle $A'B'C'$ has vertices $A' = (4, 6)$, $B' = (4, 3)$, and $C' = (-2, 3)$. Which of the transformations formed $\triangle A'B'C'$ from $\triangle ABC$?

A reflection and translation

B rotation and dilation

C reflection and rotation

D translation and rotation

58. In 1911, the population of China was 1.341403687×10^9. In the same year, the population of the United States was 3.13485438×10^8. How much greater was the population of China? For the answer to be correct, it must be in scientific notation.

F 1.027918249×10^9

G 1.654889125×10^9

H 10.27918249×10^8

J 0.1654889125×10^7

59. Look at the scatter plot below.

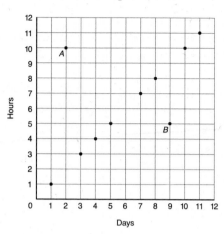

What are points *A* and *B* called?

A trend line points

B negative association points

C outliers

D scatter points

60. What is the length of the diagonal of a rectangle with vertices $A = (-5, 3)$, $B = (-5, -4)$, $C = (5, -4)$, and $D = (5, 3)$?

F $\sqrt{2}$ **H** $\sqrt{17}$

G $\sqrt{14}$ **J** Not Here

Name _____   Date _____

Practice Test B

DIRECTIONS: Read each question and choose the best answer. Use the answer sheet provided at the end of the workbook to record your answers. If the correct answer is not available, mark the letter for "Not Here."

1. Which one of the following ordered pairs does NOT define a function?

   A $(5, 5.5), (8, 7.5), (10, 8)$

   B $(-2, -6), (0, -7), (2, -8)$

   C $(-2, 4), (-1, 2), (6, 42)$

   D $(6, 2), (6, 5), (7, 21)$

2. The coordinates of triangle ABC are $A = (2, -5)$, $B = (2, -2)$, and $C = (5, -2)$. What are the coordinates of $A'B'C'$ if ABC is reflected over the x-axis?

   F $A' = (-2, -2)$, $B' = (-5, -2)$, and $C' = (-2, -5)$

   G $A' = (2, 5)$, $B' = (2, 2)$, and $C' = (5, 2)$

   H $A' = (-2, -5)$, $B' = (-2, -2)$, and $C' = (-5, -2)$

   J $A' = (-5, -2)$, $B' = (2, 5)$, and $C' = (5, 2)$

3. Which set of ordered pairs does NOT describe this linear function?

   $y = 5x - 3$

   A $(0, -3)$ and $(\frac{3}{5}, 0)$

   B $(1, 2)$ and $(\frac{4}{5}, 1)$

   C $(2, 7)$ and $(4, 18)$

   D $(5, 22)$ and $(1\frac{3}{5}, 5)$

4. Pentagons $ABCDE$ and $A'B'C'D'E'$ are drawn on a coordinate plane. $A'B'C'D'E'$ is similar but not congruent to $ABCD$. What transformation was used to create $A'B'C'D'E'$?

   F dilation

   G reflection

   H rotation

   J translation

5. In rectangle $ABCD$, $AC > AB$ and $AC \parallel BD$. What is true about these lines after a translation over 3 and up 2?

   A $A'C' > A'B'$, $A'C' \nparallel B'D'$

   B $A'C' > A'B'$, $A'C' \parallel B'D'$

   C $A'C' < A'B'$, $A'C' \parallel B'D'$

   D $A'C' < AB$, $A'C' \nparallel B'D'$

6. Look at the graph below showing two linear functions.

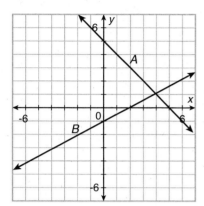

Which one of the following is true?

F Line A has a greater rate of change than line B.

G Line B has a greater rate of change than line A.

H The rates of change are equal and are equal to −1.

J The rates of change are equal and are equal to 1.

7. A ladder 10 feet tall is leaning against a building. The top of the ladder is 8 feet above the ground. How far from the building is the foot of the ladder?

A 3 feet

B 6 feet

C 8 feet

D $\sqrt{164}$ feet

8. Which one of the following is NOT a characteristic of a rational number?

F A rational number can be a repeating decimal.

G A rational number can be a nonrepeating decimal.

H A rational number can be expressed as the ratio of two integers.

J A rational number can be expressed as a fraction.

9. Rectangle ABCD is rotated 90° counterclockwise. Which one of the following is true?

A Rectangles ABCD and A′B′C′D′ are not congruent.

B Corresponding sides of ABCD and A′B′C′D′ are parallel.

C Opposite sides of A′B′C′D′ are parallel.

D The scale factor is 2.

10. Which one of the following is the equation of a line with a slope of −3 and a y-intercept of 5?

F $y = 3x + 5$

G $4y = -12x + 20$

H $2y = 6x + 10$

J $23x - 7y = 35$

11. Look at the data table and graph below showing how much two people earn for doing the same work.

Person A

Number of Windows	1	2	3	4
Earnings ($)	7	14	21	28

Person B

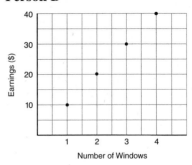

What are the rates shown, and who makes more?

A

Person A	Person B	Who earns more?
7	10	Person A

B

Person A	Person B	Who earns more?
7	10	Person B

C

Person A	Person B	Who earns more?
10	7	Person A

D

Person A	Person B	Who earns more?
10	7	Person B

12. Look at the graph below.

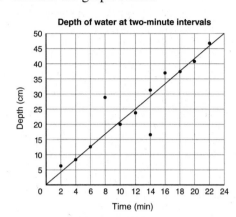

Depth of water at two-minute intervals

Which of the statements below is NOT a reason why the trend line is helpful for this graph?

F   because there are only 3 data points on the line

G   because it shows that the outliers are exceptions to the majority of the data

H   because the trend line cuts through the data points

J   because the data points are close to the trend line

---

13. The equations below define two different functions.

Function 1: $3y = 15x + 9$
Function 2: $2y = -4x + 10$

What is the rate of change of each function, and which function has the greater rate of change?

A

Function 1	Function 2	Function with Greater Rate of Change
5	−2	Function 1

B

Function 1	Function 2	Function with Greater Rate of Change
15	−4	Function 1

C

Function 1	Function 2	Function with Greater Rate of Change
15	4	Function 1

D

Function 1	Function 2	Function with Greater Rate of Change
−5	−2	Function 2

14. A cone and a square pyramid are each 25 feet high. The radius of the cone is 20 feet, and each side of the base of the pyramid is 20 feet. How do their volumes compare? Use 3.14 for π.

F The volume of the pyramid is greater than that of the cone by 466 ft³.

G The volume of the pyramid is greater than that of the cone by 7,133.33 ft³.

H The volume of the cone is greater than that of the pyramid by 7,133.33 ft³.

J The volume of the cone is greater than that of the pyramid by 466 ft³.

15. What transformation results in figures that are similar but not congruent?

A dilation

B reflection

C rotation

D translation

16. The number of tourists fluctuates by season. Tourism in summer is usually high, and then tourism is high again during the ski season. At other times, the tourist business drops and stays low. Which one of the following graphs shows this?

F

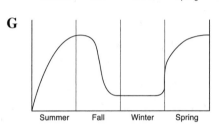

Summer   Fall   Winter   Spring

G

Summer   Fall   Winter   Spring

H

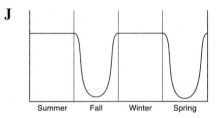

Summer   Fall   Winter   Spring

J

Summer   Fall   Winter   Spring

17. Solve the system of equations.

$y = 5x - 1$
$2y = 3x + 12$

Which ordered pair is the solution to this system of equations?

A (0, 1)   **C** (0, 6)

B $(\frac{1}{5}, 0)$ **D** (2, 9)

18. Look at the graph below.

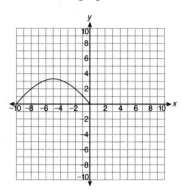

What is the average rate of change between -6 and -3?

F -2 **H** $\dfrac{1}{2}$

G $-\dfrac{1}{4}$ **J** 2

19. Look at the scatter plot below.

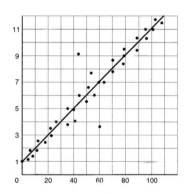

What is the rate of change and the equation of the trend line?

A rate of change: $-\dfrac{1}{10}$

   equation: $y = -\dfrac{1}{10}x + 1$

B rate of change: $\dfrac{1}{10}$

   equation: $y = \dfrac{1}{10}x + 1$

C rate of change: 10

   equation: $y = 10x + 1$

D rate of change: -10

   equation: $y = -10x + 1$

20. The amount of sugar used in a recipe is recorded in the table below.

Sugar (tsp)	1	2.5	3	4.25	7
Result (dz)	1.5	3.75	4.5	6.375	a

What is the value of a for the data to show linear growth?

F 7   **H** 12.5

G 10.5   **J** 15

21. In the figure below, $AB \parallel CD$.

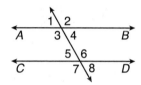

Which one of the following does NOT list pairs of equal angles?

A $\angle 1 = \angle 4,\ \angle 5 = \angle 8$

B $\angle 4 = \angle 5,\ \angle 6 = \angle 3$

C $\angle 5 = \angle 8,\ \angle 6 = \angle 7$

D $\angle 6 = \angle 4,\ \angle 3 = \angle 5$

22. The volume of a cone is 240 cm³. The radius of the base is 5 cm. How high is the cone? Use $\pi = 3.14$.

F 3.05 cm   **H** 15.28 cm

G 9.17 cm   **J** 45.88 cm

23. Triangle ABC goes through a transformation creating triangle $A'B'C'$. Which of these transformations could NOT create $A'B'C'$ from ABC such that the two triangles are congruent?

A dilation   **C** rotation

B reflection   **D** translation

24. The average distance between Mercury and Venus is about 31 million miles. What is this number expressed in scientific notation?

F 3.1×10^{-6} **G** 3.1×10^{7} **H** 3.1×10^{-7} **J** 3.1×10^{8}

---

25. What is the standard notation for 7.62953×10^{-4}?

A 0.000762953   **B** 0.00762953   **C** 7,629.53   **D** 76,295.3

---

26. Look at the system of equations.

$5x - 3y = 27$
$5x - 3y = -27$

Which one of the following best describes the solution?

F $x = 0, y = 0$ **H** all real numbers

G no solution   **J** infinite number of solutions

---

27. Rectangle $ABCD$ has vertices $A = (-5, 3)$, $B = (-5, -4)$, $C = (5, -4)$, and $D = (5, 3)$. How much longer is the diagonal of the rectangle than the longer side?

A 2.21 units   **B** 5.21 units   **C** 5.5 units   **D** 7.3 units

---

28. Look at the two tables below comparing walking speeds of two different people.

Dustin

Distance (mi)	1.5	2	4
Time (hr)	0.75	1	2

Nicolás

Distance (mi)	1	2	4
Time (hr)	0. 5	1	2

Which statement is true?

F Nicolás is faster than Dustin.   **H** Dustin and Nicolás walk at the same speed.

G Dustin is faster than Nicolás.   **J** Nicolás walks twice as fast as Dustin.

29. Trapezoid *ABCD* is rotated 90° clockwise on point *C* to form trapezoid *A′B′C′D′*. If
$A = (-5, -5)$, $B = (-5, -3)$, $C = (-1, -3)$, and $D = (-2, -5)$, what are the coordinates of *A′B′C′D′*?

A $A' = (-1, 1)$, $B' = (-3, 0)$,
   $C' = (-3, -2)$, and $D' = (-1, -3)$

B $A' = (-3, -2)$, $B' = (-1, -3)$,
   $C' = (-1, 1)$, and $D' = (3, 0)$

C $A' = (-3, 0)$, $B' = (-1, 1)$,
   $C' = (-1, -3)$, and $D' = (-3, -2)$

D $A' = (-3, 1)$, $B' = (-1, 1)$,
   $C' = (-1, -3)$, and $D' = (-3, -2)$

30. Look at the graph.

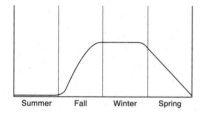

Summer   Fall   Winter   Spring

Which situation is shown in the graph?

F The greatest amount of snow falls during the winter months.

G Vegetable gardens produce mostly in the summer.

H Gas prices rise during the summer but fall during the winter.

J Gas is cheapest during the winter months.

31. Figure *ABCD* is dilated with a scale factor of 2.5. Which one of the following is true?

A The sum of the angles of *A′B′C′D′* is less than the sum of the angles of *ABCD*.

B The sum of the angles of *A′B′C′D′* is greater than the sum of the angles of *ABCD*.

C $\angle A = \angle A'$, $\angle B = \angle B'$, $\angle C = \angle C'$, and $\angle D = \angle D'$.

D The sum of angles *A*, *B*, and *C* is greater than the sum of the angles *A′*, *B′*, and *C′*.

32. Kurt is purchasing bats and baseballs for the team. The baseballs cost $20.00 each. The bats cost $85.00 each. He has up to $965.00 to spend and buys up to 19 items. How many of each can he afford?

F 9 baseballs, 10 bats

G 7 baseballs, 12 bats

H 10 baseballs, 9 bats

J Not Here

33. Solve the equation.

$$5x - 3x^2 + (2^2)^2 = 5x(1 + x) + 16 - 8x^2$$

Which one of the following best describes the solution?

A all positive numbers

B all integers

C all real numbers

D all irrational numbers

34. Look at the table below.

0	1	3	7
1	a	−5	b

What are the values for a and b that will make this a linear function?

F $a = -13, b = -1$ **G** $a = 1, b = 13$ **H** $a = -1, b = -13$ **J** $-a = 1, b = -13$

---

35. Look at the graph below.

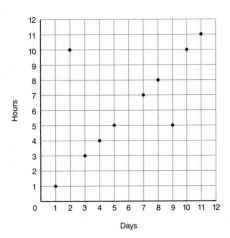

Which data chart was used to create the scatter plot?

A

1	2	3	4	5	6	7	8	9	10	11
1	2	3	4	5	6	7	8	5	10	11

B

1	2	3	4	5	6	7	8	9	10	11
1	10	3	4	5	6	7	8	5	10	11

C

1	10	3	4	5	6	7	8	5	10	11
1	2	3	4	5	6	7	8	9	10	11

D

1	2	3	4	5	6	7	8	9	10	11
1	2	3	4	5	6	7	8	9	10	11

36. Triangle *ABC* has vertices $A = (-3, 5)$, $B = (-3, -4)$, and $C = (2, -4)$. Rectangle *DEFG* has vertices $D = (-5, 3)$, $E = (-5, -4)$, $F = (5, -4)$, and $G = (5, 3)$. Which is longer, *AC* or *DF*? By how much? Round to the nearest whole number.

F *DF* is longer by 2.

G *AC* and *DF* are the same length.

H *AC* is longer by 2.

J *DF* is longer by 6.

37. If two parallel lines are reflected, what can be said about the result?

A The lines will intersect.

B The lines remain parallel, but the distance between them doubles.

C The lines remain parallel, and no other changes take place.

D The two lines merge into one line.

38. Solve the equation.

$4x(2x + 5) = 2^3x^2 + 30$

Which one of the following best describes the solution?

F $x = -1.5$

G $x = 1.5$

H There is no solution.

J There are an infinite number of solutions.

39. Look at the graph below.

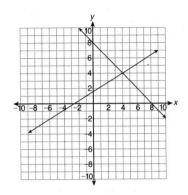

Which ordered pair is the solution to the system of equations represented by the graph?

A $(8, 8)$ **C** $(-2, 0)$

B $(4, 4)$ **D** $(-5, -2)$

40. An automobile manufacturer surveyed potential customers to see which car style they preferred. The results, shown in the table below, were categorized by men and women.

	Sedan	SUV	Total
Men	1,000	2,500	3,500
Women	2,550	1,500	4,050
Total	3,550	4,000	7,550

What is the difference in relative frequency of people who preferred sedans to SUVs?

F 6%   **H** 47%

G 20%   **J** 53%

41. Triangle *ABC* is transformed into triangle *A′B′C′* by moving it 4 across and 2 down. What is this called?

 A   a dilation

 B   a reflection

 C   a rotation

 D   a translation

42. The scatter plot and trend line go from upper left to lower right. What does this indicate about the data?

 F   The sets of data are not accurate.

 G   As one set of data increases, the other decreases.

 H   A change in one set of data does not affect the other set.

 J   Both sets of data decrease together.

43. What is $(5.73 \times 10^5)^4$ in scientific notation?

 A 1.08×10^{23}

 B 1.08×10^{20}

 C 1.08×10^9

 D 2.292×10^{21}

44. A rectangular box is 9 ft $\times$ 12 ft $\times$ 6 ft. What is the longest pipe that can be put into the box?

 F   13 ft

 G   14 ft

 H   16 ft

 J   17 ft

45. The system of equations shows the way compensation for two employees is computed where *x* represents their commission.

Employee 1: $y = 7x + 100$
Employee 2: $y = x + 700$

If both Employee 1 and Employee 2 make a commission of $50.00 during the week, which is a true statement regarding their total weekly income?

 A   Employee 2 earns more than Employee 1.

 B   Employee 1 earns more than Employee 2.

 C   Employee 1 and Employee 2 earn the same amount.

 D   Employee 1 earns $200.00 more than Employee 2.

46. Look at the figure below.

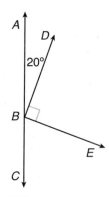

What is the measure of $\angle CBE$?

 F $\angle CBE = 180 - (90 + 20)$

 G $\angle CBE = 180 - 90 + 20$

 H $\angle CBE = 180 - \dfrac{90}{2}$

 J $\angle CBE = 180 - 90$

Name _____ Date _____

47. One half of a number equals twice the number plus three cubed. What is the number?

A -18 **C** 9

B -9 **D** 18

48. The graph below represents a function.

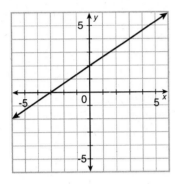

Which equation does NOT describe the function?

F $2y = -1.3x + 2$

G $y = -2x + 2$

H $3y = -2x + 6$

J $y = \frac{2}{3}x + 2$

49. Look at the scatter plot below.

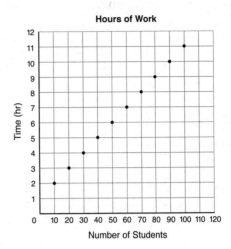

How many more students did 9 hours of homework than did 5 hours of homework?

A 30 **C** 45

B 40 **D** 50

50. Which one of the following shows these numbers ordered from least to greatest?

$$\sqrt{2},\ \sqrt[3]{8},\ \sqrt{13},\ \pi$$

F $\sqrt{2},\ \pi,\ \sqrt[3]{8},\ \sqrt{13}$

G $\sqrt{2},\ \sqrt[3]{8},\ \pi,\ \sqrt{13}$

H $\sqrt[3]{8},\ \sqrt{2},\ \pi,\ \sqrt{13}$

J $\pi,\ \sqrt{13},\ \sqrt[3]{8},\ \sqrt{2}$

51. A line goes through points $(-4, -2)$ and $(5, 0)$. A second line goes through points $(-1, 2)$ and $(4, 3)$. What is true about the solution to the system of equations represented by the lines?

A The solution consists of one point.

B The solution consists of two points.

C There are no solutions.

D There are an infinite number of solutions.

52. An auto manufacturer surveys potential customers to see which car style they prefer. The results, shown in the table below, are categorized by men and women.

	Sedan	SUV	Total
Men	1,000	2,500	3,500
Women	2,550	1,500	4,050
Total	3,550	4,000	7,550

What is the difference in relative frequency of men who like SUVs and women who like SUVs?

F 13.2%

G 19.8%

H 33.1%

J Not Here

53. Solve the equation.

$$-20 = -5x + 5$$

Which one of the following describes the solution?

A no solution

B $x = 5$

C $x = 5$ and $x = -3$

D There are an infinite number of solutions.

54. Evaluate the expression for the given values of each variable.

$$4a^2 - 6b^3 + (2a)^3 - (b^3)^2 \text{ when } a = 3 \text{ and } b = 2$$

F -70

G 18

H 140

J 172

55. Pentagon $ABCDE$ has coordinates $A = (-2, 3)$, $B = (-3, 1)$, $C = (-1, -1)$, $D = (2, -1)$, and $E = (3, 2)$. Pentagon $A'B'C'D'E'$ is formed by a dilation of $ABCDE$ and has coordinates $A' = (-5, 7.5)$, $B' = (-7.5, 2.5)$, $C' = (-2.5, -2.5)$, $D' = (5, -2.5)$, and $E' = (7.5, 5)$. What is the scale factor?

A -2.5

B -0.4

C 0.4

D 2.5

56. Look at the table below.

	Sedan	SUV
Men	a	c
Women	b	d

Which one of the following gives the correct replacements for a, b, c, and d?

F

a	b	c	d
men and sedan	women and sedan	men and SUV	women and SUV

G

a	b	c	d
women and sedan	men and sedan	men and SUV	women and SUV

H

a	b	c	d
men and sedan	women and sedan	women and SUV	men and SUV

J

a	b	c	d
women and SUV	men and SUV	women and sedan	men and sedan

---

57. How many boxes that are 6 in. × 6 in. × 6 in. will fit into a space that is 15 in. × 15 in. × 15 in.?

A 2   **B** 2.5   **C** 8   **D** 15.62

---

58. Triangle ABC is reflected over the y-axis, forming triangle $A'B'C'$. Which one of the following is true?

F $\angle A = \angle A', \angle B = \angle B', \angle C > \angle C'$

G $\angle A = \angle A', \angle B = \angle B', \angle C = \angle C'$

H $\angle A + \angle B + \angle C < \angle A' + \angle B' + \angle C'$

J $\angle A + \angle B + \angle C > \angle A' + \angle B' + \angle C'$

59. The smallest country in the world is the Vatican, with an area of 1 km². The largest country is Russia, with an area of 17,075,200 km². How many times larger than the Vatican is Russia?

   A 1.7075200×10^{-7}

   B 1.7075200×10^{8}

   C 1.7075200×10^{-6}

   D 1.7075200×10^{7}

60. In which one of the following is a trend line NOT helpful?

   F when the data points are linear

   G when the data points show no association

   H when the data points cluster around the trend line

   J when the data points show negative association

Answer Sheets

Pretest

1 Ⓐ Ⓑ Ⓒ Ⓓ	11 Ⓐ Ⓑ Ⓒ Ⓓ	21 Ⓐ Ⓑ Ⓒ Ⓓ	31 Ⓐ Ⓑ Ⓒ Ⓓ	41 Ⓐ Ⓑ Ⓒ Ⓓ	51 Ⓐ Ⓑ Ⓒ Ⓓ
2 Ⓕ Ⓖ Ⓗ Ⓙ	12 Ⓕ Ⓖ Ⓗ Ⓙ	22 Ⓕ Ⓖ Ⓗ Ⓙ	32 Ⓕ Ⓖ Ⓗ Ⓙ	42 Ⓕ Ⓖ Ⓗ Ⓙ	52 Ⓕ Ⓖ Ⓗ Ⓙ
3 Ⓐ Ⓑ Ⓒ Ⓓ	13 Ⓐ Ⓑ Ⓒ Ⓓ	23 Ⓐ Ⓑ Ⓒ Ⓓ	33 Ⓐ Ⓑ Ⓒ Ⓓ	43 Ⓐ Ⓑ Ⓒ Ⓓ	53 Ⓐ Ⓑ Ⓒ Ⓓ
4 Ⓕ Ⓖ Ⓗ Ⓙ	14 Ⓕ Ⓖ Ⓗ Ⓙ	24 Ⓕ Ⓖ Ⓗ Ⓙ	34 Ⓕ Ⓖ Ⓗ Ⓙ	44 Ⓕ Ⓖ Ⓗ Ⓙ	54 Ⓕ Ⓖ Ⓗ Ⓙ
5 Ⓐ Ⓑ Ⓒ Ⓓ	15 Ⓐ Ⓑ Ⓒ Ⓓ	25 Ⓐ Ⓑ Ⓒ Ⓓ	35 Ⓐ Ⓑ Ⓒ Ⓓ	45 Ⓐ Ⓑ Ⓒ Ⓓ	55 Ⓐ Ⓑ Ⓒ Ⓓ
6 Ⓕ Ⓖ Ⓗ Ⓙ	16 Ⓕ Ⓖ Ⓗ Ⓙ	26 Ⓕ Ⓖ Ⓗ Ⓙ	36 Ⓕ Ⓖ Ⓗ Ⓙ	46 Ⓕ Ⓖ Ⓗ Ⓙ	56 Ⓕ Ⓖ Ⓗ Ⓙ
7 Ⓐ Ⓑ Ⓒ Ⓓ	17 Ⓐ Ⓑ Ⓒ Ⓓ	27 Ⓐ Ⓑ Ⓒ Ⓓ	37 Ⓐ Ⓑ Ⓒ Ⓓ	47 Ⓐ Ⓑ Ⓒ Ⓓ	57 Ⓐ Ⓑ Ⓒ Ⓓ
8 Ⓕ Ⓖ Ⓗ Ⓙ	18 Ⓕ Ⓖ Ⓗ Ⓙ	28 Ⓕ Ⓖ Ⓗ Ⓙ	38 Ⓕ Ⓖ Ⓗ Ⓙ	48 Ⓕ Ⓖ Ⓗ Ⓙ	58 Ⓕ Ⓖ Ⓗ Ⓙ
9 Ⓐ Ⓑ Ⓒ Ⓓ	19 Ⓐ Ⓑ Ⓒ Ⓓ	29 Ⓐ Ⓑ Ⓒ Ⓓ	39 Ⓐ Ⓑ Ⓒ Ⓓ	49 Ⓐ Ⓑ Ⓒ Ⓓ	59 Ⓐ Ⓑ Ⓒ Ⓓ
10 Ⓕ Ⓖ Ⓗ Ⓙ	20 Ⓕ Ⓖ Ⓗ Ⓙ	30 Ⓕ Ⓖ Ⓗ Ⓙ	40 Ⓕ Ⓖ Ⓗ Ⓙ	50 Ⓕ Ⓖ Ⓗ Ⓙ	60 Ⓕ Ⓖ Ⓗ Ⓙ

The Number System Modeled Instruction

1 Ⓐ Ⓑ Ⓒ Ⓓ	2 Ⓕ Ⓖ Ⓗ Ⓙ	3 Ⓐ Ⓑ Ⓒ Ⓓ	4 Ⓕ Ⓖ Ⓗ Ⓙ	5 Ⓐ Ⓑ Ⓒ Ⓓ	6 Ⓕ Ⓖ Ⓗ Ⓙ

The Number System Independent Practice

7 Ⓐ Ⓑ Ⓒ Ⓓ	10 Ⓕ Ⓖ Ⓗ Ⓙ	13 Ⓐ Ⓑ Ⓒ Ⓓ	16 Ⓕ Ⓖ Ⓗ Ⓙ	19 Ⓐ Ⓑ Ⓒ Ⓓ	
8 Ⓕ Ⓖ Ⓗ Ⓙ	11 Ⓐ Ⓑ Ⓒ Ⓓ	14 Ⓕ Ⓖ Ⓗ Ⓙ	17 Ⓐ Ⓑ Ⓒ Ⓓ	20 Ⓕ Ⓖ Ⓗ Ⓙ	
9 Ⓐ Ⓑ Ⓒ Ⓓ	12 Ⓕ Ⓖ Ⓗ Ⓙ	15 Ⓐ Ⓑ Ⓒ Ⓓ	18 Ⓕ Ⓖ Ⓗ Ⓙ		

Expressions and Equations Modeled Instruction

1 Ⓐ Ⓑ Ⓒ Ⓓ	10 Ⓕ Ⓖ Ⓗ Ⓙ	19 Ⓐ Ⓑ Ⓒ Ⓓ	28 Ⓕ Ⓖ Ⓗ Ⓙ	37 Ⓐ Ⓑ Ⓒ Ⓓ	46 Ⓕ Ⓖ Ⓗ Ⓙ
2 Ⓕ Ⓖ Ⓗ Ⓙ	11 Ⓐ Ⓑ Ⓒ Ⓓ	20 Ⓕ Ⓖ Ⓗ Ⓙ	29 Ⓐ Ⓑ Ⓒ Ⓓ	38 Ⓕ Ⓖ Ⓗ Ⓙ	47 Ⓐ Ⓑ Ⓒ Ⓓ
3 Ⓐ Ⓑ Ⓒ Ⓓ	12 Ⓕ Ⓖ Ⓗ Ⓙ	21 Ⓐ Ⓑ Ⓒ Ⓓ	30 Ⓕ Ⓖ Ⓗ Ⓙ	39 Ⓐ Ⓑ Ⓒ Ⓓ	48 Ⓕ Ⓖ Ⓗ Ⓙ
4 Ⓕ Ⓖ Ⓗ Ⓙ	13 Ⓐ Ⓑ Ⓒ Ⓓ	22 Ⓕ Ⓖ Ⓗ Ⓙ	31 Ⓐ Ⓑ Ⓒ Ⓓ	40 Ⓕ Ⓖ Ⓗ Ⓙ	49 Ⓐ Ⓑ Ⓒ Ⓓ
5 Ⓐ Ⓑ Ⓒ Ⓓ	14 Ⓕ Ⓖ Ⓗ Ⓙ	23 Ⓐ Ⓑ Ⓒ Ⓓ	32 Ⓕ Ⓖ Ⓗ Ⓙ	41 Ⓐ Ⓑ Ⓒ Ⓓ	50 Ⓕ Ⓖ Ⓗ Ⓙ
6 Ⓕ Ⓖ Ⓗ Ⓙ	15 Ⓐ Ⓑ Ⓒ Ⓓ	24 Ⓕ Ⓖ Ⓗ Ⓙ	33 Ⓐ Ⓑ Ⓒ Ⓓ	42 Ⓕ Ⓖ Ⓗ Ⓙ	
7 Ⓐ Ⓑ Ⓒ Ⓓ	16 Ⓕ Ⓖ Ⓗ Ⓙ	25 Ⓐ Ⓑ Ⓒ Ⓓ	34 Ⓕ Ⓖ Ⓗ Ⓙ	43 Ⓐ Ⓑ Ⓒ Ⓓ	
8 Ⓕ Ⓖ Ⓗ Ⓙ	17 Ⓐ Ⓑ Ⓒ Ⓓ	26 Ⓕ Ⓖ Ⓗ Ⓙ	35 Ⓐ Ⓑ Ⓒ Ⓓ	44 Ⓕ Ⓖ Ⓗ Ⓙ	
9 Ⓐ Ⓑ Ⓒ Ⓓ	18 Ⓕ Ⓖ Ⓗ Ⓙ	27 Ⓐ Ⓑ Ⓒ Ⓓ	36 Ⓕ Ⓖ Ⓗ Ⓙ	45 Ⓐ Ⓑ Ⓒ Ⓓ	

Expressions and Equations Independent Practice

51 Ⓐ Ⓑ Ⓒ Ⓓ	61 Ⓐ Ⓑ Ⓒ Ⓓ	71 Ⓐ Ⓑ Ⓒ Ⓓ	81 Ⓐ Ⓑ Ⓒ Ⓓ	91 Ⓐ Ⓑ Ⓒ Ⓓ	101 Ⓐ Ⓑ Ⓒ Ⓓ
52 Ⓕ Ⓖ Ⓗ Ⓙ	62 Ⓕ Ⓖ Ⓗ Ⓙ	72 Ⓕ Ⓖ Ⓗ Ⓙ	82 Ⓕ Ⓖ Ⓗ Ⓙ	92 Ⓕ Ⓖ Ⓗ Ⓙ	102 Ⓕ Ⓖ Ⓗ Ⓙ
53 Ⓐ Ⓑ Ⓒ Ⓓ	63 Ⓐ Ⓑ Ⓒ Ⓓ	73 Ⓐ Ⓑ Ⓒ Ⓓ	83 Ⓐ Ⓑ Ⓒ Ⓓ	93 Ⓐ Ⓑ Ⓒ Ⓓ	103 Ⓐ Ⓑ Ⓒ Ⓓ
54 Ⓕ Ⓖ Ⓗ Ⓙ	64 Ⓕ Ⓖ Ⓗ Ⓙ	74 Ⓕ Ⓖ Ⓗ Ⓙ	84 Ⓕ Ⓖ Ⓗ Ⓙ	94 Ⓕ Ⓖ Ⓗ Ⓙ	104 Ⓕ Ⓖ Ⓗ Ⓙ
55 Ⓐ Ⓑ Ⓒ Ⓓ	65 Ⓐ Ⓑ Ⓒ Ⓓ	75 Ⓐ Ⓑ Ⓒ Ⓓ	85 Ⓐ Ⓑ Ⓒ Ⓓ	95 Ⓐ Ⓑ Ⓒ Ⓓ	
56 Ⓕ Ⓖ Ⓗ Ⓙ	66 Ⓕ Ⓖ Ⓗ Ⓙ	76 Ⓕ Ⓖ Ⓗ Ⓙ	86 Ⓕ Ⓖ Ⓗ Ⓙ	96 Ⓕ Ⓖ Ⓗ Ⓙ	
57 Ⓐ Ⓑ Ⓒ Ⓓ	67 Ⓐ Ⓑ Ⓒ Ⓓ	77 Ⓐ Ⓑ Ⓒ Ⓓ	87 Ⓐ Ⓑ Ⓒ Ⓓ	97 Ⓐ Ⓑ Ⓒ Ⓓ	
58 Ⓕ Ⓖ Ⓗ Ⓙ	68 Ⓕ Ⓖ Ⓗ Ⓙ	78 Ⓕ Ⓖ Ⓗ Ⓙ	88 Ⓕ Ⓖ Ⓗ Ⓙ	98 Ⓕ Ⓖ Ⓗ Ⓙ	
59 Ⓐ Ⓑ Ⓒ Ⓓ	69 Ⓐ Ⓑ Ⓒ Ⓓ	79 Ⓐ Ⓑ Ⓒ Ⓓ	89 Ⓐ Ⓑ Ⓒ Ⓓ	99 Ⓐ Ⓑ Ⓒ Ⓓ	
60 Ⓕ Ⓖ Ⓗ Ⓙ	70 Ⓕ Ⓖ Ⓗ Ⓙ	80 Ⓕ Ⓖ Ⓗ Ⓙ	90 Ⓕ Ⓖ Ⓗ Ⓙ	100 Ⓕ Ⓖ Ⓗ Ⓙ	

Functions Modeled Instruction

1 Ⓐ Ⓑ Ⓒ Ⓓ   3 Ⓐ Ⓑ Ⓒ Ⓓ   5 Ⓐ Ⓑ Ⓒ Ⓓ   7 Ⓐ Ⓑ Ⓒ Ⓓ   9 Ⓐ Ⓑ Ⓒ Ⓓ
2 Ⓕ Ⓖ Ⓗ Ⓙ   4 Ⓕ Ⓖ Ⓗ Ⓙ   6 Ⓕ Ⓖ Ⓗ Ⓙ   8 Ⓕ Ⓖ Ⓗ Ⓙ   10 Ⓕ Ⓖ Ⓗ Ⓙ

Functions Independent Practice

11 Ⓐ Ⓑ Ⓒ Ⓓ   15 Ⓐ Ⓑ Ⓒ Ⓓ   19 Ⓐ Ⓑ Ⓒ Ⓓ   23 Ⓐ Ⓑ Ⓒ Ⓓ   27 Ⓐ Ⓑ Ⓒ Ⓓ
12 Ⓕ Ⓖ Ⓗ Ⓙ   16 Ⓕ Ⓖ Ⓗ Ⓙ   20 Ⓕ Ⓖ Ⓗ Ⓙ   24 Ⓕ Ⓖ Ⓗ Ⓙ   28 Ⓕ Ⓖ Ⓗ Ⓙ
13 Ⓐ Ⓑ Ⓒ Ⓓ   17 Ⓐ Ⓑ Ⓒ Ⓓ   21 Ⓐ Ⓑ Ⓒ Ⓓ   25 Ⓐ Ⓑ Ⓒ Ⓓ   29 Ⓐ Ⓑ Ⓒ Ⓓ
14 Ⓕ Ⓖ Ⓗ Ⓙ   18 Ⓕ Ⓖ Ⓗ Ⓙ   22 Ⓕ Ⓖ Ⓗ Ⓙ   26 Ⓕ Ⓖ Ⓗ Ⓙ   30 Ⓕ Ⓖ Ⓗ Ⓙ

Geometry Modeled Instruction

1 Ⓐ Ⓑ Ⓒ Ⓓ   10 Ⓕ Ⓖ Ⓗ Ⓙ   19 Ⓐ Ⓑ Ⓒ Ⓓ   28 Ⓕ Ⓖ Ⓗ Ⓙ   37 Ⓐ Ⓑ Ⓒ Ⓓ   46 Ⓕ Ⓖ Ⓗ Ⓙ
2 Ⓕ Ⓖ Ⓗ Ⓙ   11 Ⓐ Ⓑ Ⓒ Ⓓ   20 Ⓕ Ⓖ Ⓗ Ⓙ   29 Ⓐ Ⓑ Ⓒ Ⓓ   38 Ⓕ Ⓖ Ⓗ Ⓙ   47 Ⓐ Ⓑ Ⓒ Ⓓ
3 Ⓐ Ⓑ Ⓒ Ⓓ   12 Ⓕ Ⓖ Ⓗ Ⓙ   21 Ⓐ Ⓑ Ⓒ Ⓓ   30 Ⓕ Ⓖ Ⓗ Ⓙ   39 Ⓐ Ⓑ Ⓒ Ⓓ   48 Ⓕ Ⓖ Ⓗ Ⓙ
4 Ⓕ Ⓖ Ⓗ Ⓙ   13 Ⓐ Ⓑ Ⓒ Ⓓ   22 Ⓕ Ⓖ Ⓗ Ⓙ   31 Ⓐ Ⓑ Ⓒ Ⓓ   40 Ⓕ Ⓖ Ⓗ Ⓙ   49 Ⓐ Ⓑ Ⓒ Ⓓ
5 Ⓐ Ⓑ Ⓒ Ⓓ   14 Ⓕ Ⓖ Ⓗ Ⓙ   23 Ⓐ Ⓑ Ⓒ Ⓓ   32 Ⓕ Ⓖ Ⓗ Ⓙ   41 Ⓐ Ⓑ Ⓒ Ⓓ   50 Ⓕ Ⓖ Ⓗ Ⓙ
6 Ⓕ Ⓖ Ⓗ Ⓙ   15 Ⓐ Ⓑ Ⓒ Ⓓ   24 Ⓕ Ⓖ Ⓗ Ⓙ   33 Ⓐ Ⓑ Ⓒ Ⓓ   42 Ⓕ Ⓖ Ⓗ Ⓙ
7 Ⓐ Ⓑ Ⓒ Ⓓ   16 Ⓕ Ⓖ Ⓗ Ⓙ   25 Ⓐ Ⓑ Ⓒ Ⓓ   34 Ⓕ Ⓖ Ⓗ Ⓙ   43 Ⓐ Ⓑ Ⓒ Ⓓ
8 Ⓕ Ⓖ Ⓗ Ⓙ   17 Ⓐ Ⓑ Ⓒ Ⓓ   26 Ⓕ Ⓖ Ⓗ Ⓙ   35 Ⓐ Ⓑ Ⓒ Ⓓ   44 Ⓕ Ⓖ Ⓗ Ⓙ
9 Ⓐ Ⓑ Ⓒ Ⓓ   18 Ⓕ Ⓖ Ⓗ Ⓙ   27 Ⓐ Ⓑ Ⓒ Ⓓ   36 Ⓕ Ⓖ Ⓗ Ⓙ   45 Ⓐ Ⓑ Ⓒ Ⓓ

Geometry Independent Practice

51 Ⓐ Ⓑ Ⓒ Ⓓ   62 Ⓕ Ⓖ Ⓗ Ⓙ   73 Ⓐ Ⓑ Ⓒ Ⓓ   84 Ⓕ Ⓖ Ⓗ Ⓙ   95 Ⓐ Ⓑ Ⓒ Ⓓ   106 Ⓕ Ⓖ Ⓗ Ⓙ
52 Ⓕ Ⓖ Ⓗ Ⓙ   63 Ⓐ Ⓑ Ⓒ Ⓓ   74 Ⓕ Ⓖ Ⓗ Ⓙ   85 Ⓐ Ⓑ Ⓒ Ⓓ   96 Ⓕ Ⓖ Ⓗ Ⓙ   107 Ⓐ Ⓑ Ⓒ Ⓓ
53 Ⓐ Ⓑ Ⓒ Ⓓ   64 Ⓕ Ⓖ Ⓗ Ⓙ   75 Ⓐ Ⓑ Ⓒ Ⓓ   86 Ⓕ Ⓖ Ⓗ Ⓙ   97 Ⓐ Ⓑ Ⓒ Ⓓ   108 Ⓕ Ⓖ Ⓗ Ⓙ
54 Ⓕ Ⓖ Ⓗ Ⓙ   65 Ⓐ Ⓑ Ⓒ Ⓓ   76 Ⓕ Ⓖ Ⓗ Ⓙ   87 Ⓐ Ⓑ Ⓒ Ⓓ   98 Ⓕ Ⓖ Ⓗ Ⓙ   109 Ⓐ Ⓑ Ⓒ Ⓓ
55 Ⓐ Ⓑ Ⓒ Ⓓ   66 Ⓕ Ⓖ Ⓗ Ⓙ   77 Ⓐ Ⓑ Ⓒ Ⓓ   88 Ⓕ Ⓖ Ⓗ Ⓙ   99 Ⓐ Ⓑ Ⓒ Ⓓ   110 Ⓕ Ⓖ Ⓗ Ⓙ
56 Ⓕ Ⓖ Ⓗ Ⓙ   67 Ⓐ Ⓑ Ⓒ Ⓓ   78 Ⓕ Ⓖ Ⓗ Ⓙ   89 Ⓐ Ⓑ Ⓒ Ⓓ   100 Ⓕ Ⓖ Ⓗ Ⓙ
57 Ⓐ Ⓑ Ⓒ Ⓓ   68 Ⓕ Ⓖ Ⓗ Ⓙ   79 Ⓐ Ⓑ Ⓒ Ⓓ   90 Ⓕ Ⓖ Ⓗ Ⓙ   101 Ⓐ Ⓑ Ⓒ Ⓓ
58 Ⓕ Ⓖ Ⓗ Ⓙ   69 Ⓐ Ⓑ Ⓒ Ⓓ   80 Ⓕ Ⓖ Ⓗ Ⓙ   91 Ⓐ Ⓑ Ⓒ Ⓓ   102 Ⓕ Ⓖ Ⓗ Ⓙ
59 Ⓐ Ⓑ Ⓒ Ⓓ   70 Ⓕ Ⓖ Ⓗ Ⓙ   81 Ⓐ Ⓑ Ⓒ Ⓓ   92 Ⓕ Ⓖ Ⓗ Ⓙ   103 Ⓐ Ⓑ Ⓒ Ⓓ
60 Ⓕ Ⓖ Ⓗ Ⓙ   71 Ⓐ Ⓑ Ⓒ Ⓓ   82 Ⓕ Ⓖ Ⓗ Ⓙ   93 Ⓐ Ⓑ Ⓒ Ⓓ   104 Ⓕ Ⓖ Ⓗ Ⓙ
61 Ⓐ Ⓑ Ⓒ Ⓓ   72 Ⓕ Ⓖ Ⓗ Ⓙ   83 Ⓐ Ⓑ Ⓒ Ⓓ   94 Ⓕ Ⓖ Ⓗ Ⓙ   105 Ⓐ Ⓑ Ⓒ Ⓓ

Statistics and Probability Modeled Instruction

1 Ⓐ Ⓑ Ⓒ Ⓓ   3 Ⓐ Ⓑ Ⓒ Ⓓ   5 Ⓐ Ⓑ Ⓒ Ⓓ   7 Ⓐ Ⓑ Ⓒ Ⓓ
2 Ⓕ Ⓖ Ⓗ Ⓙ   4 Ⓕ Ⓖ Ⓗ Ⓙ   6 Ⓕ Ⓖ Ⓗ Ⓙ   8 Ⓕ Ⓖ Ⓗ Ⓙ

Statistics and Probability Independent Practice

9 Ⓐ Ⓑ Ⓒ Ⓓ   12 Ⓕ Ⓖ Ⓗ Ⓙ   15 Ⓐ Ⓑ Ⓒ Ⓓ   18 Ⓕ Ⓖ Ⓗ Ⓙ   21 Ⓐ Ⓑ Ⓒ Ⓓ   24 Ⓕ Ⓖ Ⓗ Ⓙ
10 Ⓕ Ⓖ Ⓗ Ⓙ   13 Ⓐ Ⓑ Ⓒ Ⓓ   16 Ⓕ Ⓖ Ⓗ Ⓙ   19 Ⓐ Ⓑ Ⓒ Ⓓ   22 Ⓕ Ⓖ Ⓗ Ⓙ
11 Ⓐ Ⓑ Ⓒ Ⓓ   14 Ⓕ Ⓖ Ⓗ Ⓙ   17 Ⓐ Ⓑ Ⓒ Ⓓ   20 Ⓕ Ⓖ Ⓗ Ⓙ   23 Ⓐ Ⓑ Ⓒ Ⓓ

Answer Sheets
Higher Scores on Math, Grade 8

Practice Test A

1 Ⓐ Ⓑ Ⓒ Ⓓ	11 Ⓐ Ⓑ Ⓒ Ⓓ	21 Ⓐ Ⓑ Ⓒ Ⓓ	31 Ⓐ Ⓑ Ⓒ Ⓓ	41 Ⓐ Ⓑ Ⓒ Ⓓ	51 Ⓐ Ⓑ Ⓒ Ⓓ
2 Ⓕ Ⓖ Ⓗ Ⓙ	12 Ⓕ Ⓖ Ⓗ Ⓙ	22 Ⓕ Ⓖ Ⓗ Ⓙ	32 Ⓕ Ⓖ Ⓗ Ⓙ	42 Ⓕ Ⓖ Ⓗ Ⓙ	52 Ⓕ Ⓖ Ⓗ Ⓙ
3 Ⓐ Ⓑ Ⓒ Ⓓ	13 Ⓐ Ⓑ Ⓒ Ⓓ	23 Ⓐ Ⓑ Ⓒ Ⓓ	33 Ⓐ Ⓑ Ⓒ Ⓓ	43 Ⓐ Ⓑ Ⓒ Ⓓ	53 Ⓐ Ⓑ Ⓒ Ⓓ
4 Ⓕ Ⓖ Ⓗ Ⓙ	14 Ⓕ Ⓖ Ⓗ Ⓙ	24 Ⓕ Ⓖ Ⓗ Ⓙ	34 Ⓕ Ⓖ Ⓗ Ⓙ	44 Ⓕ Ⓖ Ⓗ Ⓙ	54 Ⓕ Ⓖ Ⓗ Ⓙ
5 Ⓐ Ⓑ Ⓒ Ⓓ	15 Ⓐ Ⓑ Ⓒ Ⓓ	25 Ⓐ Ⓑ Ⓒ Ⓓ	35 Ⓐ Ⓑ Ⓒ Ⓓ	45 Ⓐ Ⓑ Ⓒ Ⓓ	55 Ⓐ Ⓑ Ⓒ Ⓓ
6 Ⓕ Ⓖ Ⓗ Ⓙ	16 Ⓕ Ⓖ Ⓗ Ⓙ	26 Ⓕ Ⓖ Ⓗ Ⓙ	36 Ⓕ Ⓖ Ⓗ Ⓙ	46 Ⓕ Ⓖ Ⓗ Ⓙ	56 Ⓕ Ⓖ Ⓗ Ⓙ
7 Ⓐ Ⓑ Ⓒ Ⓓ	17 Ⓐ Ⓑ Ⓒ Ⓓ	27 Ⓐ Ⓑ Ⓒ Ⓓ	37 Ⓐ Ⓑ Ⓒ Ⓓ	47 Ⓐ Ⓑ Ⓒ Ⓓ	57 Ⓐ Ⓑ Ⓒ Ⓓ
8 Ⓕ Ⓖ Ⓗ Ⓙ	18 Ⓕ Ⓖ Ⓗ Ⓙ	28 Ⓕ Ⓖ Ⓗ Ⓙ	38 Ⓕ Ⓖ Ⓗ Ⓙ	48 Ⓕ Ⓖ Ⓗ Ⓙ	58 Ⓕ Ⓖ Ⓗ Ⓙ
9 Ⓐ Ⓑ Ⓒ Ⓓ	19 Ⓐ Ⓑ Ⓒ Ⓓ	29 Ⓐ Ⓑ Ⓒ Ⓓ	39 Ⓐ Ⓑ Ⓒ Ⓓ	49 Ⓐ Ⓑ Ⓒ Ⓓ	59 Ⓐ Ⓑ Ⓒ Ⓓ
10 Ⓕ Ⓖ Ⓗ Ⓙ	20 Ⓕ Ⓖ Ⓗ Ⓙ	30 Ⓕ Ⓖ Ⓗ Ⓙ	40 Ⓕ Ⓖ Ⓗ Ⓙ	50 Ⓕ Ⓖ Ⓗ Ⓙ	60 Ⓕ Ⓖ Ⓗ Ⓙ

Practice Test B

1 Ⓐ Ⓑ Ⓒ Ⓓ	11 Ⓐ Ⓑ Ⓒ Ⓓ	21 Ⓐ Ⓑ Ⓒ Ⓓ	31 Ⓐ Ⓑ Ⓒ Ⓓ	41 Ⓐ Ⓑ Ⓒ Ⓓ	51 Ⓐ Ⓑ Ⓒ Ⓓ
2 Ⓕ Ⓖ Ⓗ Ⓙ	12 Ⓕ Ⓖ Ⓗ Ⓙ	22 Ⓕ Ⓖ Ⓗ Ⓙ	32 Ⓕ Ⓖ Ⓗ Ⓙ	42 Ⓕ Ⓖ Ⓗ Ⓙ	52 Ⓕ Ⓖ Ⓗ Ⓙ
3 Ⓐ Ⓑ Ⓒ Ⓓ	13 Ⓐ Ⓑ Ⓒ Ⓓ	23 Ⓐ Ⓑ Ⓒ Ⓓ	33 Ⓐ Ⓑ Ⓒ Ⓓ	43 Ⓐ Ⓑ Ⓒ Ⓓ	53 Ⓐ Ⓑ Ⓒ Ⓓ
4 Ⓕ Ⓖ Ⓗ Ⓙ	14 Ⓕ Ⓖ Ⓗ Ⓙ	24 Ⓕ Ⓖ Ⓗ Ⓙ	34 Ⓕ Ⓖ Ⓗ Ⓙ	44 Ⓕ Ⓖ Ⓗ Ⓙ	54 Ⓕ Ⓖ Ⓗ Ⓙ
5 Ⓐ Ⓑ Ⓒ Ⓓ	15 Ⓐ Ⓑ Ⓒ Ⓓ	25 Ⓐ Ⓑ Ⓒ Ⓓ	35 Ⓐ Ⓑ Ⓒ Ⓓ	45 Ⓐ Ⓑ Ⓒ Ⓓ	55 Ⓐ Ⓑ Ⓒ Ⓓ
6 Ⓕ Ⓖ Ⓗ Ⓙ	16 Ⓕ Ⓖ Ⓗ Ⓙ	26 Ⓕ Ⓖ Ⓗ Ⓙ	36 Ⓕ Ⓖ Ⓗ Ⓙ	46 Ⓕ Ⓖ Ⓗ Ⓙ	56 Ⓕ Ⓖ Ⓗ Ⓙ
7 Ⓐ Ⓑ Ⓒ Ⓓ	17 Ⓐ Ⓑ Ⓒ Ⓓ	27 Ⓐ Ⓑ Ⓒ Ⓓ	37 Ⓐ Ⓑ Ⓒ Ⓓ	47 Ⓐ Ⓑ Ⓒ Ⓓ	57 Ⓐ Ⓑ Ⓒ Ⓓ
8 Ⓕ Ⓖ Ⓗ Ⓙ	18 Ⓕ Ⓖ Ⓗ Ⓙ	28 Ⓕ Ⓖ Ⓗ Ⓙ	38 Ⓕ Ⓖ Ⓗ Ⓙ	48 Ⓕ Ⓖ Ⓗ Ⓙ	58 Ⓕ Ⓖ Ⓗ Ⓙ
9 Ⓐ Ⓑ Ⓒ Ⓓ	19 Ⓐ Ⓑ Ⓒ Ⓓ	29 Ⓐ Ⓑ Ⓒ Ⓓ	39 Ⓐ Ⓑ Ⓒ Ⓓ	49 Ⓐ Ⓑ Ⓒ Ⓓ	59 Ⓐ Ⓑ Ⓒ Ⓓ
10 Ⓕ Ⓖ Ⓗ Ⓙ	20 Ⓕ Ⓖ Ⓗ Ⓙ	30 Ⓕ Ⓖ Ⓗ Ⓙ	40 Ⓕ Ⓖ Ⓗ Ⓙ	50 Ⓕ Ⓖ Ⓗ Ⓙ	60 Ⓕ Ⓖ Ⓗ Ⓙ

Answer Key

Pretest

1. B [8.NS.1]
2. J [8.NS.2]
3. C [8.NS.1]
4. H [8.NS.2]
5. B [8.EE.1]
6. F [8.EE.2]
7. B [8.EE.3]
8. H [8.EE.4]
9. B [8.EE.5]
10. F [8.EE.6]
11. A [8.EE.7.a]
12. F [8.EE.7.b]
13. B [8.EE.8.a]
14. H [8.EE.8.b]
15. C [8.EE.8.c]
16. H [8.EE.5]
17. D [8.EE.6]
18. G [8.EE.7.a]
19. A [8.EE.7.b]
20. H [8.EE.8.a]
21. B [8.EE.8.b]
22. J [8.EE.8.c]
23. A [8.F.1]
24. F [8.F.2]
25. B [8.F.3]
26. H [8.F.4]
27. B [8.F.5]
28. H [8.F.1]
29. B [8.F.2]
30. G [8.F.3]
31. A [8.F.4]
32. H [8.F.5]
33. D [8.G.1.a]
34. J [8.G.1.b]
35. B [8.G.1.c]

36. G [8.G.2]
37. B [8.G.3]
38. F [8.G.4]
39. C [8.G.5]
40. G [8.G.7]
41. D [8.G.7]
42. H [8.G.8]
43. B [8.G.9]
44. G [8.G.1.a]
45. C [8.G.1.b]
46. F [8.G.1.c]
47. A [8.G.2]
48. F [8.G.3]
49. A [8.G.4]
50. G [8.G.5]
51. C [8.SP.1]
52. F [8.SP.2]
53. B [8.SP.3]
54. H [8.SP.4]
55. B [8.SP.1]
56. G [8.SP.2]
57. A [8.SP.3]
58. G [8.SP.4]
59. D [8.SP.1]
60. F [8.SP.4]

The Number System
Modeled Instruction

1. B [8.NS.1]
2. J [8.NS.2]
3. C [8.NS.1]
4. J [8.NS.2]
5. A [8.NS.1]
6. F [8.NS.2]

The Number System
Independent Practice

7. D [8.NS.2]
8. G [8.NS.2]
9. C [8.NS.2]
10. F [8.NS.1]
11. A [8.NS.2]
12. H [8.NS.1]
13. A [8.NS.1]
14. F [8.NS.2]
15. C [8.NS.2]
16. G [8.NS.2]
17. B [8.NS.1]
18. H [8.NS.1]
19. C [8.NS.1]
20. G [8.NS.1]

Expressions and Equations
Modeled Instruction

1. C [8.EE.5]
2. H [8.EE.1]
3. D [8.EE.2]
4. F [8.EE.6]
5. C [8.EE.8.a]
6. H [8.EE.8.b]
7. D [8.EE.7.b]
8. F [8.EE.7.a]
9. B [8.EE.8.c]
10. G [8.EE.4]
11. C [8.EE.6]
12. H [8.EE.3]
13. C [8.EE.8.a]
14. J [8.EE.6]
15. C [8.EE.6]
16. H [8.EE.8.b]
17. A [8.EE.7.a]
18. J [8.EE.7.b]

Answer Key
Higher Scores on Math, Grade 8

19. A [8.EE.7.b]
20. F [8.EE.7.a]
21. B [8.EE.3]
22. F [8.EE.5]
23. D [8.EE.5]
24. H [8.EE.8.c]
25. B [8.EE.2]
26. G [8.EE.1]
27. C [8.EE.4]
28. H [8.EE.4]
29. C [8.EE.4]
30. H [8.EE.5]
31. A [8.EE.3]
32. H [8.EE.3]
33. B [8.EE.2]
34. H [8.EE.1]
35. B [8.EE.8.c]
36. H [8.EE.7.b]
37. D [8.EE.8.b]
38. G [8.EE.8.a]
39. B [8.EE.8.a]
40. F [8.EE.8.c]
41. D [8.EE.1]
42. J [8.EE.7.a]
43. D [8.EE.8.a]
44. F [8.EE.7.a]
45. B [8.EE.2]
46. G [8.EE.8.c]
47. B [8.EE.8.b]
48. H [8.EE.8.b]
49. A [8.EE.6]
50. G [8.EE.7.b]

Expressions and Equations
Independent Practice

51. B [8.EE.2]
52. F [8.EE.4]
53. A [8.EE.7.a]
54. H [8.EE.7.b]
55. A [8.EE.8.c]

56. G [8.EE.7.a]
57. A [8.EE.8.c]
58. F [8.EE.4]
59. D [8.EE.1]
60. F [8.EE.7.a]
61. C [8.EE.8.b]
62. J [8.EE.5]
63. B [8.EE.3]
64. G [8.EE.6]
65. C [8.EE.8.b]
66. J [8.EE.5]
67. D [8.EE.7.a]
68. G [8.EE.1]
69. B [8.EE.5]
70. F [8.EE.3]
71. D [8.EE.6]
72. G [8.EE.8.b]
73. D [8.EE.7.b]
74. F [8.EE.7.b]
75. C [8.EE.8.b]
76. F [8.EE.3]
77. C [8.EE.8.b]
78. G [8.EE.4]
79. D [8.EE.7.b]
80. H [8.EE.8.b]
81. B [8.EE.7.a]
82. G [8.EE.7.a]
83. B [8.EE.7.a]
84. H [8.EE.6]
85. D [8.EE.2]
86. F [8.EE.8.c]
87. D [8.EE.5]
88. H [8.EE.4]
89. B [8.EE.8.c]
90. F [8.EE.2]
91. D [8.EE.8.a]
92. G [8.EE.6]
93. A [8.EE.1]
94. G [8.EE.6]
95. C [8.EE.8.a]

96. J [8.EE.4]
97. C [8.EE.5]
98. G [8.EE.5]
99. B [8.EE.6]
100. G [8.EE.8.a]
101. A [8.EE.1]
102. G [8.EE.7.b]
103. A [8.EE.3]
104. J [8.EE.8.a]

Functions
Modeled Instruction

1. C [8.F.1]
2. F [8.F.2]
3. B [8.F.3]
4. F [8.F.4]
5. B [8.F.5]
6. H [8.F.1]
7. C [8.F.2]
8. H [8.F.3]
9. A [8.F.4]
10. J [8.F.5]

Functions
Independent Practice

11. B [8.F.5]
12. F [8.F.3]
13. B [8.F.4]
14. G [8.F.4]
15. A [8.F.3]
16. G [8.F.2]
17. C [8.F.3]
18. F [8.F.4]
19. C [8.F.1]
20. J [8.F.4]
21. B [8.F.2]
22. G [8.F.1]
23. B [8.F.5]
24. H [8.F.3]
25. D [8.F.1]

26. G [8.F.2]

27. D [8.F.5]

28. F [8.F.5]

29. D [8.F.2]

30. G [8.F.1]

Geometry
Modeled Instruction

1. A [8.G.1.a]

2. G [8.G.1.c]

3. B [8.G.9]

4. J [8.G.4]

5. C [8.G.1.c]

6. F [8.G.4]

7. C [8.G.2]

8. F [8.G.3]

9. B [8.G.9]

10. H [8.G.3]

11. C [8.G.7]

12. J [8.G.9]

13. B [8.G.1.b]

14. H [8.G.8]

15. A [8.G.9]

16. H [8.G.8]

17. B [8.G.8]

18. H [8.G.1.c]

19. B [8.G.7]

20. H [8.G.7]

21. C [8.G.9]

22. F [8.G.1.b]

23. D [8.G.1.a]

24. F [8.G.5]

25. A [8.G.3]

26. J [8.G.4]

27. B [8.G.3]

28. J [8.G.1.c]

29. B [8.G.7]

30. J [8.G.7]

31. A [8.G.1.a]

32. F [8.G.5]

33. A [8.G.9]

34. F [8.G.1.c]

35. D [8.G.1.b]

36. J [8.G.2]

37. B [8.G.2]

38. J [8.G.1.b]

39. B [8.G.4]

40. F [8.G.4]

41. C [8.G.5]

42. J [8.G.7]

43. B [8.G.1.a]

44. G [8.G.1.a]

45. A [8.G.5]

46. F [8.G.4]

47. C [8.G.1.b]

48. J [8.G.5]

49. C [8.G.2]

50. H [8.G.2]

Geometry
Independent Practice

51. C [8.G.1.a]

52. J [8.G.8]

53. D [8.G.1.c]

54. H [8.G.8]

55. B [8.G.3]

56. G [8.G.1.b]

57. C [8.G.7]

58. H [8.G.7]

59. C [8.G.5]

60. F [8.G.1.c]

61. B [8.G.1.b]

62. F [8.G.4]

63. D [8.G.8]

64. J [8.G.1.b]

65. B [8.G.1.a]

66. F [8.G.2]

67. B [8.G.2]

68. H [8.G.9]

69. C [8.G.1.b]

70. G [8.G.2]

71. C [8.G.1.a]

72. H [8.G.5]

73. D [8.G.3]

74. F [8.G.1.c]

75. D [8.G.7]

76. H [8.G.4]

77. A [8.G.4]

78. G [8.G.1.a]

79. D [8.G.9]

80. H [8.G.2]

81. B [8.G.1.c]

82. J [8.G.8]

83. C [8.G.4]

84. J [8.G.7]

85. B [8.G.9]

86. J [8.G.1.a]

87. B [8.G.3]

88. H [8.G.7]

89. C [8.G.7]

90. J [8.G.1.b]

91. A [8.G.4]

92. F [8.G.7]

93. C [8.G.1.c]

94. F [8.G.1.c]

95. D [8.G.5]

96. F [8.G.2]

97. B [8.G.1.b]

98. H [8.G.9]

99. B [8.G.3]

100. H [8.G.3]

101. B [8.G.1.b]

102. H [8.G.1.b]

103. A [8.G.3]

104. J [8.G.8]

105. D [8.G.4]

106. J [8.G.5]

107. C [8.G.9]

108. G [8.G.3]

109. D [8.G.8]

110. F [8.G.5]

Statistics and Probability

Modeled Instruction

1. B [8.SP.1]
2. H [8.SP.2]
3. A [8.SP.3]
4. G [8.SP.4]
5. A [8.SP.1]
6. H [8.SP.3]
7. B [8.SP.3]
8. H [8.SP.4]

Statistics and Probability

Independent Practice

9. C [8.SP.4]
10. J [8.SP.2]
11. A [8.SP.3]
12. H [8.SP.3]
13. C [8.SP.3]
14. H [8.SP.1]
15. D [8.SP.4]
16. F [8.SP.4]
17. B [8.SP.3]
18. J [8.SP.4]
19. B [8.SP.2]
20. F [8.SP.1]
21. B [8.SP.3]
22. G [8.SP.1]
23. C [8.SP.2]
24. J [8.SP.1]

Practice Test A

1. B [8.G.5]
2. F [8.EE.2]
3. C [8.G.1.a]
4. F [8.EE.4]
5. A [8.SP.3]
6. H [8.G.4]
7. A [8.SP.3]
8. H [8.F.2]

9. C [8.EE.6]
10. H [8.SP.2]
11. C [8.NS.1]
12. G [8.G.1.a]
13. C [8.G.1.b]
14. H [8.G.1.c]
15. C [8.F.5]
16. G [8.EE.1]
17. A [8.SP.4]
18. H [8.EE.7.b]
19. A [8.G.9]
20. H [8.G.3]
21. B [8.EE.5]
22. H [8.NS.2]
23. C [8.EE.8.b]
24. F [8.F.1]
25. B [8.F.3]
26. F [8.EE.6]
27. C [8.SP.1]
28. G [8.G.5]
29. C [8.G.7]
30. F [8.EE.8.b]
31. B [8.G.1.c]
32. F [8.G.7]
33. C [8.EE.8.c]
34. F [8.EE.8.c]
35. A [8.EE.7.a]
36. G [8.G.9]
37. C [8.F.4]
38. H [8.F.4]
39. B [8.EE.8.a]
40. J [8.G.1.b]
41. D [8.EE.8.a]
42. H [8.G.8]
43. A [8.G.2]
44. J [8.F.3]
45. C [8.SP.2]
46. J [8.EE.7.b]
47. B [8.EE.5]
48. H [8.F.5]

49. C [8.EE.7.a]
50. F [8.F.2]
51. D [8.EE.4]
52. J [8.G.3]
53. B [8.EE.3]
54. F [8.G.4]
55. B [8.F.1]
56. G [8.SP.4]
57. B [8.G.2]
58. F [8.EE.3]
59. C [8.SP.1]
60. J [8.G.8]

Practice Test B

1. D [8.F.1]
2. G [8.G.3]
3. C [8.F.3]
4. F [8.G.4]
5. B [8.G.1.a]
6. G [8.F.4]
7. B [8.G.7]
8. G [8.NS.1]
9. C [8.G.1.c]
10. G [8.EE.6]
11. B [8.EE.5]
12. F [8.SP.2]
13. A [8.F.2]
14. H [8.G.9]
15. A [8.G.2]
16. J [8.F.5]
17. D [8.EE.8.b]
18. G [8.EE.5]
19. B [8.SP.3]
20. G [8.EE.6]
21. D [8.G.5]
22. G [8.G.9]
23. A [8.G.2]
24. G [8.EE.3]
25. A [8.EE.3]
26. G [8.EE.8.b]

27. A [8.G.8]

28. H [8.F.2]

29. D [8.G.4]

30. F [8.F.5]

31. C [8.G.1.b]

32. H [8.EE.8.c]

33. C [8.EE.7.a]

34. H [8.F.1]

35. B [8.SP.1]

36. F [8.G.8]

37. C [8.G.1.a]

38. G [8.EE.7.b]

39. B [8.EE.8.a]

40. F [8.SP.4]

41. D [8.G.1.c]

42. G [8.SP.1]

43. A [8.EE.4]

44. H [8.G.7]

45. A [8.F.4]

46. F [8.G.5]

47. A [8.EE.7.b]

48. J [8.F.3]

49. B [8.SP.3]

50. G [8.NS.2]

51. A [8.EE.8.c]

52. F [8.SP.4]

53. B [8.EE.7.a]

54. H [8.EE.1]

55. D [8.G.3]

56. F [8.EE.8.a]

57. C [8.EE.2]

58. G [8.G.1.b]

59. D [8.EE.4]

60. G [8.SP.2]

Reference Sheet

Length

Customary

1 mile (mi) = 1,760 yards (yd)

1 yard (yd) = 3 feet (ft)

1 foot (ft) = 12 inches (in.)

Metric

1 kilometer (km) = 1,000 meters (m)

1 meter (m) = 100 centimeters (cm)

1 centimeter (cm) = 10 millimeters (mm)

Volume and Capacity

Customary

1 gallon (gal) = 4 quarts (qt)

1 quart (qt) = 2 pints (pt)

1 pint (pt) = 2 cups (c)

1 cup (c) = 8 fluid ounces (fl oz)

Metric

1 liter (L) = 1,000 milliliters (mL)

Weight and Mass

Customary

1 ton (T) = 2,000 pounds (lb)

1 pound (lb) = 16 ounces (oz)

Metric

1 kilogram (kg) = 1,000 grams (g)

1 gram (g) = 1,000 milligrams (mg)

Perimeter of a Rectangle	$P = 2l + 2w$

Circumference	$C = 2\pi r$ or $C = \pi d$

Area

Triangle	$A = \frac{1}{2}bh$
Rectangle	$A = bh$
Parallelogram	$A = bh$
Trapezoid	$A = \frac{1}{2}(b_1 + b_2)h$
Circle	$A = \pi r^2$

Volume

Prism	$V = Bh$
Pyramid	$V = \frac{1}{3}Bh$
Cylinder	$V = \pi r^2 h$ or $V = Bh$
Sphere	$V = \frac{4}{3}\pi r^3$
Cone	$V = \frac{1}{3}\pi r^2 h$ or $V = \frac{1}{3}Bh$

Slope	$\dfrac{y_2 - y_1}{x_2 - x_1}$

Pythagorean Theorem	$a^2 + b^2 = c^2$